COST CONTROL IN THE
CONSTRUCTION INDUSTRY

COST CONTROL IN THE CONSTRUCTION INDUSTRY

J. GOBOURNE, AIOB, AMBIM, MICPS
Area Development Engineer, Sir Alfred McAlpine and Son Ltd

LONDON
NEWNES-BUTTERWORTHS

THE BUTTERWORTH GROUP
ENGLAND
Butterworth & Co (Publishers) Ltd
London: 88 Kingsway, WC2B 6AB

AUSTRALIA
Butterworths Pty Ltd
Sydney: 586 Pacific Highway, NSW 2067
Melbourne: 343 Little Collins Street, 3000
Brisbane: 240 Queen Street, 4000

CANADA
Butterworth & Co (Canada) Ltd
Toronto: 14 Curity Avenue, 374

NEW ZEALAND
Butterworths of New Zealand Ltd
Wellington: 26–28 Waring Taylor Street, 1

SOUTH AFRICA
Butterworth & Co (South Africa) (Pty) Ltd
Durban: 152–154 Gale Street

First published in 1973 by Newnes-Butterworths
an imprint of the Butterworth Group

© J. Gobourne, 1973

ISBN 0 408 00087 2 Standard
     0 408 00088 0 Limp

Printed in England by Chapel River Press, Andover, Hants.

PREFACE

In this book I have described the principles of cost control as applicable to the construction industry, and have shown how these principles can be applied to measuring and controlling the utilisation of labour, plant materials and overheads on a construction site.

No two companies are likely to evolve absolutely identical costing systems, because they are constantly governed by factors outside the costing system itself, such as accountancy procedures, payroll systems, methods of estimating, degree of planning, incentive schemes, use of computer, staff availability, company policies and types and density of construction undertaken. For this reason the book continually refers to variations of the point under discussion in order that the reader, upon completion, can tailor-make a cost control system to suit his own company's conditions and to incorporate whatever additional factors may be required by that company, e.g. incentive bonus payments, feed-back of output data to estimators, profitability of various trades, ratios of tradesmen to labourers, exclusion of overheads and inclusion of staff.

The examples of cost control systems given in the book, although primarily intended for Institute of Building Associate examination students, can be seen to apply equally to many aspects of site construction work, such as civil, mechanical and electrical, the variation being one of technical phraseology rather than basic principles.

J. Gobourne

Wolverhampton, 1972

PUBLISHERS' NOTE

Names of personnel given in the various task sheets, time allocation sheets, etc., throughout this book, are completely imaginary and bear no relationship to actual persons living or dead.

ACKNOWLEDGEMENT

Acknowledgement is due to the Construction Industry Consul of the Universal Esperanto Association in Stockholm, Inĝeniero per Törnegren Fakdelegito de U.E.A. por Konstrutekniko, who provided such useful information, both in Esperanto and in English, on the piece rate system and work study data bank of the Federation of Svedish Building Employers.

CONTENTS

1	Introduction to Cost Control	1
2	Contract Accounts and Prime Costs	5
3	Unit Costing	11
4	Standard Costing	18
5	Expenditure Data for Unit Costing and Standard Costing	27
6	Standards for Measurable Work	40
7	Standards for Site Overheads	52
8	Feed-back of Output Data	58
9	Incorporating a Bonus System	61
10	Standard Cost Example	65
11	Standard Cost Exercise	99
12	Costing Sundry Items	115
13	Materials	130
14	Sub-Contractors	135
15	Action	137
	Index	147

1
INTRODUCTION TO COST CONTROL

FIRM'S ACCOUNTS

A firm's annual profit and loss account provides the master control of all the firm's activities. If an unsatisfactory balance is shown year after year, it is plain that action is required to prevent eventual bankruptcy. One would think that the problem would never be allowed to go that far; yet the building and civil engineering industry is noted for its high rate of bankruptcies, not only among the smaller builders but also among the larger companies.

CONTRACT ACCOUNTS

By separating accounts for individual contracts, both of money earned and money spent, an account can be built up of the profitability of each individual contract. Costs not applicable to any one particular contract, such as head office overheads, directors' fees, bank charges, taxation, etc., can be apportioned to each contract as, for example, a percentage of the final certified value of the contract, a cost per capita, or a percentage of the tender figure or quotation.

PRIME COSTS

The majority of contracts are of such duration that interim payments are sanctioned by the client during construction. The frequency of these payments will depend upon the client's specification for minimum applications, but is usually once a month.
  Provided the interim payment is a reliable reflection of the work

carried out so far, a comparison can be made between this payment and the expenditure incurred to the same date; this comparison provides an interim picture of the site's profitability.

SITE COSTING

The degree of control that can be achieved by a prime cost is, of course, limited because of its over-all nature. Although a prime cost may expose a frightening loss on a contract, it does not pinpoint that loss or indicate its nature. Examples are: loss being made by carpenters on retaining wall structure; loss being made on cement owing to over-rich concrete; loss being made on S.E.T. owing to insufficient overtime being worked.

To show this kind of detail the contract must have a systematic, regular and continuous check of all the various elements that go to make up the earnings and expenditure of that contract.

The contract's earnings are almost exclusively confined to the amount being paid by the client for carrying out the contract and fall into the following categories.

1. Measured work paid for at rates agreed between the client and the contractor, either by tender or by negotiation.
2. Daywork based on rates similarly agreed between the client and the contractor.*
3. Preliminaries paid for as a percentage or lump sum on the contract figure and similarly agreed between the client and contractor.
4. Nominated materials and nominated sub-contractors paid for at cost to the contractor, plus an agreed percentage to cover the contractor's obligations of attendance, etc.
5. Increases in costs where the contract entitles the contractor to recover such increases. These are based on a list of basic prices, which the contractor will have quoted as his basics when tendering.

---

* For reference: *Schedule of Basic Plant Charges for Use in Connection with Dayworks under a Building Contract*, price 50p, from The Royal Institution of Chartered Surveyors, 12 Great George Street, Parliament Square, London, S.W.1. *The Federation of Civil Engineering Contractors Schedules of Dayworks Carried out Incidental to Contract Work*, price 25p, from Romney House, Tufton Street, Westminster, London, S.W.1.

Introduction to Cost Control

Earnings not paid for by the client are, for example:

1. Sundry sales of surplus or scrap materials or plant.
2. Sundry works for other clients.
3. Tax, S.E.T. or rate refunds.

The contract's expenditure is extremely diverse but can be grouped into the following categories.

1. The direct cost of labour paid in the form of wages or piece-work payments to the contractor's own labour force and to labour-only sub-contractors.
2. Plant charges, including running costs such as fuel, oil, spares, etc.
3. Site overheads such as temporary roads, offices, etc., and cost of holiday, insurance stamps, travelling expenses, etc.
4. Invoices for materials, both permanent materials such as bricks or cement and temporary materials such as shuttering, sheet piles or scaffolding.
5. Sub-contractor's accounts based on measured rates or daywork agreed with the sub-contractor.
6. A proportion of the contractor's head office establishment charges.

This diversity of earnings and expenditure necessitates differing approaches to the various elements to be controlled, but these elements can basically be divided into the following two groups.

Those requiring regular control

1. Labour costs.
2. Plant charges.
3. Site overheads.

These can best be controlled by a system of unit costing or standard costing at a weekly or even daily frequency.

Those which do not normally need to be reviewed regularly

1. Materials.
2. Sub-contractors.
3. Head Office charges.

Except for the continuous wastage factor which occurs with the material element and the attendance factor which occurs with sub-contractors, the necessity to control this group is normally confined to the initial calculation of:

1. Buying margins.
2. Sub-letting margins.
3. Head Office budgeted percentage.

SPOT COSTS

A spot cost is a cost check carried out on site on an individual element or section of the contract. Spot costs may be made for many reasons, the three main ones being as follows.

1. As a check on the regular site cost control system. Usually this is carried out by a detailed observation of a random operation or section of work being made and compared with the results shown on the regular system.
2. As a detailed study of a particular operation or section of work where the regular site cost control system is not used or does not give the required information in the correct form or detail.
3. As a recorded cost of a particular operation agreed with the client or the client's representative as being the actual cost of an operation and used in later negotiation as the basis for a revision of payment.

2
CONTRACT ACCOUNTS AND PRIME COSTS

Both contract accounts and prime costs require the same type of information, i.e. the total of all earnings and of all expenditure and the comparison of these totals for a contract—at the end of the contract for contract accounts or at intervals during the contract for prime costs.

Example 2.1. Contract accounts summary

Earnings	£
Measured work, including variations and P.C. materials	103 467·90
Dayworks	921·05
Preliminaries	750·00
Nominated sub-contractors	12 325·73
	£117 464·68

Expenditure	£
Workmen's wages	41 228·37
Site staff wages	3 420·91
Labour-only sub-contractors	846·50
Internal plant charges	11 242·06
External plant charges	725·63
Invoices for materials	30 287·82
Payments to sub-contractors (including nominated)	20 423·25
Sundry site expenses	591·62
Head Office costs—$2\frac{1}{2}\%$ of final certificate, i.e. $2\frac{1}{2}\%$ of £117 465	2 936·62
	£111 702·78

	£
Total earnings	117 464·68
Total expenditure	111 702·78
Profit on contract	£5 761·90

$= 5·2\%$ on expenditure

An additional problem that arises with prime costs is the possible earnings to the contract of contentious payments such as revised rates to be negotiated owing to some revision by the client, claims to be discussed or dayworks to be agreed. These items may or may not be paid to the contractor, and it is wise therefore to exclude them from any prime cost calculations showing their *possible* increase to the earnings of the contract after the financial comparison has been made. It is possible that a large contract or even a small problematical contract may carry a heavy sum of contentious items, in which case the amount of contention could be defined under the headings of, for example, 'probable', 'possible' or 'unlikely' payments by the client and a number of differing totals for the contract earnings be obtained. No single formula can be calculated for evaluating contentious sums, and each contract must be considered separately when large contentions are involved. However, the fact remains that, regardless of the amount of the contention, the true earnings of a contract must not include that contention in the initial comparison.

Retention, however, is a different matter, as the client is obliged eventually to pay it to the contractor upon completion of the works. Retention need not therefore be deducted from the earnings in the comparison on the assumption that the contractor will, in fact, complete the contract.

Example 2.2. Prime cost summary

Earnings	£
Measured work, including variations and P.C. materials	26 150
Dayworks authorised to date	220
Preliminaries	450
Materials on site	200
Nominated sub-contractors	—
	£27 020

Expenditure	£
Workmen's wages	15 025
Site staff wages	1 200
Labour only sub-contractors	450
Internal plant charges	1 043
External plant charges	185
Invoices for materials	6 830
Payments to sub-contractors	715
Sundry site expenses	124
Head Office costs—$2\frac{1}{2}\%$ of interim certificate, i.e. $2\frac{1}{2}\%$ of £27 020	675
	£26 247

Contract Accounts and Prime Costs

Total earnings	27 020
Total expenditure	26 247
Profit on contract	£773

$= 2 \cdot 95\%$ on expenditure

Contentious items	£
Dayworks outstanding	53
Variations rejected	28
Possible additional profit	£81
Total possible profit	£854

$= 3 \cdot 25\%$ on expenditure

If the earnings have been loosely calculated to provide a quick and approximate interim valuation, any comparison between earnings and expenditure can only have a limited credibility. By setting a maximum variation either way to the valuation it may be possible to make limited use of such a figure.

Example 2.3

Total earnings	£123 500
Possible variation between limits of $+3$ and -5%	
Therefore maximum value of earnings	£127 205
minimum value of earnings	£117 325
Total expenditure	£128 135

Therefore loss between limits of £930 and £10 810 $= 0 \cdot 73$ to $8 \cdot 45\%$ on expenditure.

Programmed Earnings

By cashing out the bar lines or activities of the contract programme at bill of quantity rates, the programmed earnings of the contract can be calculated; these shown as a table or as a graph can be compared with actual earnings and will indicate the financial progress of a contract, as indicated in *Figure 2.1*.

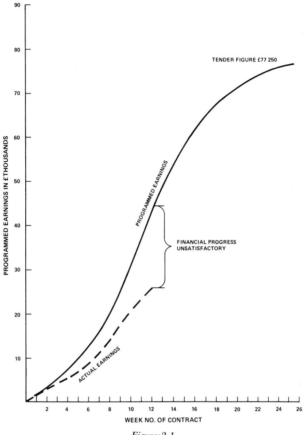

Figure 2.1

Prime Cost Breakdown

By applying percentages to the earnings side of a prime cost it is possible to compare the various elements of expenditure with the same elements of earnings. Rule of thumb, statistics from previous contracts or even a complete programmed breakdown of bill rates can be used to calculate these percentages to an acceptable degree of accuracy.

Example 2.2, page 7 shows total earnings of £27 020 which may be broken down as follows:

Contract Accounts and Prime Costs

Earnings	Labour		Plant		Materials		Sub-Contrs.		H.O. Costs	
	%	£	%	£	%	£	%	£	%	£
Material on site £26 820	60	16 090	15	4 020	20	200 5 360	2½	675	2½	675
£27 020						5 560				

The expenditure is already broken down:

Expenditure	Labour	Plant	Materials	Sub-Contrs.	H.O. Costs
	£	£	£	£	£
Workmen's wages	15 025				
Site staff wages	1 200				
Lab. only contractors	450				
Internal plant charges		1 043			
External plant charges		185			
Invoices for materials			6 830		
Sub-contractors				715	
Sundry site expenses	124				
Head office costs					675
Total £26 247	16 799	1 228	6 830	715	675

Comparison of earnings and expenditure shows the following:

Comparison	Labour	Plant	Materials	Sub-Contrs.	H.O. Costs
	£	£	£	£	
Gains	—	2 792	—	—	—
Losses	709	—	1 265	40	—

This indicates a loss somewhere on the material element and the possibility of work being carried out by hand when machine methods had been envisaged at tender.

Exercise

Calculate a prime cost based on the following information.

	£
Measured work to date	240 390
Variations authorised to date	34 806
Variations in contention	5 723

Nominated sub-contractors	21 240
Dayworks authorised to date	8 916
Dayworks outstanding	721
V.O.P. payments authorised	1 920
Materials on site	400
Preliminaries	7 600
Claim on preliminaries for extended contract	1 800
Labour on loan to other sites	751
Labour on loan from other sites	73
Workmen's wages	103 256
Site staff wages	5 634
Invoices for materials	64 220
Sundry site expenses	920
External plant charges	8 220
Internal plant charges	16 575
Sub-contractors (including nominated)	89 372
Head Office costs—4% of site expenditure	

Group Studies

1. Discuss the various problems encountered in obtaining the information necessary to calculate a prime cost.

2. Discuss the various methods of charging a contract with head office overheads and advise on the most suitable method for companies carrying out various types of work.

3
UNIT COSTING

The basis of unit costing is the computation of expenditure either in cash or in hours of an operation per single unit of measurement.

Example 3.1

If 6 men take 4 h to hand dig an interceptor pit of size 3·5 m × 2·0 m 1·0 m deep, i.e. 7 m³, then the unit cost in hours of digging the pit is:

$$\frac{6 \times 4}{7} = \underline{3\cdot 4 \text{ h/m}^3}$$

Example 3.2

Assuming a basic labour rate of $37\frac{1}{2}$p/h, the unit cost in cash would be:

$$\frac{6 \times 4 \times 37\frac{1}{2}\text{p}}{7} = \underline{\pounds 1\cdot 29/\text{m}^3 \text{ nett}}$$

This unit cost is nett because it has been cashed out at the *basic* labour rate and does not include overheads of any kind.

A gross unit cost would be obtained by calculating first the gross cost of a man-hour. This is achieved by dividing the cash total of the week's wages sheets by the number of hours worked. For instance, *Figure 10.15* shows a wage sheet summary; the gross cost of an average man-hour for that week is:

$$\frac{\pounds 325\cdot 61}{620\frac{3}{4} \text{ h}} = \underline{52\frac{1}{2}\text{p/h}}$$

Example 3.3

The gross unit of cost of excavating the interceptor pit during that week would therefore be:

$$\frac{6 \times 4 \times 52\frac{1}{2}\text{p}}{7} = \underline{\pounds 1\cdot 80/\text{m}^3 \text{ gross}}$$

Unit Costing

W/E 19/2/72

WEEK NO. 3   SHEET NO. 1

M	T	W	T	F	S	S	TOTAL HOURS	MEASURE	DESCRIPTION	UNIT COST
10	15		10	20			55	12 m³	Excavate bases by hand	4.6 h/m³
18 ⑨	18 ⑨	18 ⑨	10 ④				64 ㉛	425 m³	Excavate basement by 22 RB (plant)	6.7 m³/h 13.7 m³/h
			8	15			23	3 m³	Concrete column bases	7.6 h/m³
9							9	2 m³	Blinding	4.5 h/m³
	7	18	21				46	7 m²	Fix shutters to columns	6.6 h/m²
18	9		6				33	6 m²	Make beam shutters	5.5 h/m²
				18	15		33	28 m²	Strip base shutters	1.2 h/m²
				9	8		17	8 m²	Make wall shutters	2.1 h/m²
18	18	18	18	18	15		105	45 m²	Brickwork to manholes 1B thick	2.3 h/m²

Figure 3.1

Figure 3.1 shows a sample sheet from a unit cost statement, the hours being the man-hours spent on each operation day by day.

Operations involving the use of plant can be shown in a different colour to labour and the unit cost calculated in man-hours per unit, plus plant-hours per unit.

The conversion of the hours per operation into cash, either nett or gross, makes it possible to add together the labour and plant elements of an operation, whereas, when unit costs are presented in hours, this addition is only possible if plant-hours are converted into the equivalent man-hours, e.g. 1 No. 22RB hour = 3·75 man-hours. Thus the operation of 'excavating basement by 22RB' illustrated in *Figure 3.1* could be converted as shown in *Figure 3.2*.

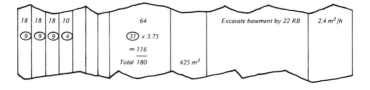

Figure 3.2

Unit Costing

A typical unit cost form showing cash results is illustrated in *Figure 3.3*.

Unit costing provides immediate feedback of output data for use in future tenders; however, the regularity required for site cost control, usually weekly, is far too frequent for the contractor's

M	T	W	T	F	S	S	TOTAL HOURS	@ 37½p = £ COST	MEASURE	DESCRIPTION	UNIT COST
10	15		10	20			55	20.62	12 m^3	Excavate bases by hand	£1.72/m^3

Figure 3.3

estimators to revise their knowledge of outputs. Conditions on site, holidays, accidents, location, sickness, weather, etc., may vary a unit cost considerably from week to week, and only long-term information can be of use in the contractor's output library.

Unit costing, although more straightforward to operate than standard costing, has the following disadvantages.

1. The system does not highlight the major losses, and without some means of comparison, either with previous weeks or with a library of average outputs, the problem of spotting the losing items is left to the more obvious operations where outputs are well known. The introduction of the metric system into the industry increases this problem, as years of well-learnt outputs have to be unlearnt for new metric units.
2. The over-all profit or loss of the contract is not known for the period of the unit cost. By addition the over-all expenditure could be obtained but this would not indicate whether that expenditure were in excess of the contract's earnings for the period or otherwise.
3. Although overheads can be included in the rate for calculation of a gross unit cost, there is no way of controlling the overheads as a separate problem. For example, £1·80 per cubic metre gross may be a good output for excavating an interceptor pit under particular site conditions, but perhaps the rate could be better if, for instance, fewer men were being paid lodging allowance. Without further investigation this question cannot be answered in a unit costing system.
4. On the purely mathematical side there are two points against unit costing: first, the vast amount of divisional calculation required, division being extremely prone to human error;

						THIS WEEK				TO DATE		
M	T	W	T	F	S	S	TOTAL HOURS	MEASURE	DESCRIPTION	UNIT COST h/m^2	UNIT COST h/m^2	MEASURE m^2	TOTAL HOURS
	7	18	21				46	$7\,m^2$	Fix shutters to columns	6.6	—	—	—
18	9		6				33	$6\,m^2$	Make beam shutters	5.5	4.2	110	460
				18	15		33	$28\,m^2$	Strip base shutters	1.2	1.2	212	253
				9	8		17	$8\,m^2$	Make wall shutters	2.1	2.2	28	61

Figure 3.4

Unit Costing

and second, the fact that calculations cannot be balanced as a check against errors. Standard cost statements can be balanced with the wage sheets, plant books, etc., to check that total expenditure is correct.

USE OF UNIT COSTING AS A CONTROL TOOL

Three basic methods are available for controlling costs by a unit cost system.

1. From a basic unit cost statement as shown in *Figures 3.1–3.3* site management can compare unit costs from week to week, watching for changes and studying further the items that suddenly show an increased unit cost. The statement can be extended to show the previous week's unit cost against each operation, which makes this comparison less difficult.
2. By comparing the week's unit cost for an operation with the 'to-date' unit cost for that operation, management can see whether the week's production is above or below the site's normal output. The reason for a sudden change can be researched and the appropriate action taken either to return production to normal or to record the interference causing the change and possibly recover the additional costs from the client. *Figure 3.4* illustrates a weekly and cumulative unit cost statement. It can be seen that a comparison of the week's unit cost and the to-date unit cost quickly indicates which operations require attention. An even more effective comparison is that between this week's unit costs and the to-date unit costs of the previous week, i.e. before the to-date figures have been affected by this week's results.

Example 3.4

The previous to-date records for an item of brickwork to retaining wall are as in *Figure 3.5*.

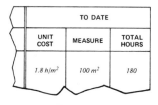

Figure 3.5

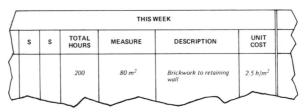

Figure 3.6

	TO DATE	
UNIT COST	MEASURE	TOTAL HOURS
2.1 h/m²	180 m²	380

Figure 3.7

W/E 19/2/72   SHEET NO. 1

WEEK NO. 3

M	T	W	T	F	S	S	TOTAL HOURS	MEASURE	DESCRIPTION	UNIT COST	UNIT STANDARD
10	15		10	20			55	12 m³	Exc. bases by hand	4.6 h/m³	4.0
18	18	18	10				64	425 m³	Exc. basement by 22 RB (plant)	6.7 m³/h (13.7 m³/h)	7.5 (15.0)
⑨	⑨	⑨	④				㉛				
			8		15		23	3 m³	Concrete column bases	7.6 h/m³	6.0
9							9	2 m³	Blinding	4.5 h/m³	7.0
	7	18	21				46	7 m²	Fix shutters to cols.	6.6 h/m²	7.0
18	9		6				33	6 m²	Make beam shutters	5.5 h/m²	4.5
				18	15		33	28 m²	Strip base shutters	1.2 h/m²	1.0
				9	8		17	8 m²	Make wall shutters	2.1 h/m²	2.2
18	18	18	18	18	15		105	45 m²	Brickwork to manhole. 1B thick	2.3 h/m²	2.5

Figure 3.8

Unit Costing

If this week's results are as in *Figure 3.6*, then the new to-date figures will be as in *Figure 3.7*.

The drop in output from the previous total of $1\cdot8$ h/m² to this week's $2\cdot5$ h/m² is more pronounced and more likely to arouse action than the drop within the week from an average of $2\cdot1$ h/m² to $2\cdot5$ h/m².

3. By setting a target or standard output as being the norm for each operation and comparing the unit cost with this standard, the operations showing a loss on standard can be picked out and studied further to find the reasons for their losses. *Figure 3.8* illustrates a unit cost statement with standards included.

4
STANDARD COSTING

DEFINITION OF STANDARD COSTING

The Institute of Cost and Works Accountants* defines the basic principles of standard costing as:

1. the establishment of a predetermined standard or target of performance;
2. the measurement of actual performance;
3. the comparison of actual performance, in detail and in total, with the predetermined standard;
4. the disclosure of variances between actual and standard performance and the reasons for these variances;
5. the suggestion of corrective action where examination of the variances indicates that this is necessary.

By common usage the application of these principles to the detailed production operations and products is termed 'standard costing'.

In short, standard costing is the computation of expenditure, either in cash or in hours of an operation, and the comparison of that expenditure with a known standard or value for that amount of work.

Example 4.1

If 6 men take 4 h to hand dig an interceptor pit of size $3.5 \text{ m} \times 2.0 \text{ m} \times 1.0 \text{ m}$ deep, i.e. 7 m^3, then the cost of digging that pit is:

$$6 \times 4 = \underline{24 \text{ h cost}}$$

---

* *An Introduction to Budgetary Control, Standard Costing, Material Control and Production Control*, price 75p, from The Institute of Cost and Works Accountants, 63 Portland Place, London, W.1.

Standard Costing

If the contractor's standard value for carrying out this kind of work under similar conditions is 3·5 h/m³, then the total standard value for this pit is:

$$7 \times 3 \cdot 5 = \underline{24 \cdot 5 \text{ h value}}$$

Comparing the cost with the standard value

Value	24·5
Cost	24·0
Gain on operation	0·5 h

As with unit costing, these results can be expressed either in nett costs or in gross costs, i.e. excluding overheads or including overheads.

Example 4.2

Nett cash value	
$7 \times 3 \cdot 5 \times 37\frac{1}{2}$p	= £9·19
Nett cash cost $6 \times 4 \times 37\frac{1}{2}$p	= £9·00
Gain on operation (nett)	£0·19

Example 4.3

Gross cash value	
$7 \times 3 \cdot 5 \times 52\frac{1}{2}$p	= £12·86
Gross cash cost $6 \times 4 \times 52\frac{1}{2}$p	= £12·60
Gain on operation (gross)	£ 0·26

Figure 4.1 shows a sample sheet from a standard cost statement, the hours being the man-hours spent on each operation day by day.

Site management is left in no doubt as to which operations are below standard, as they are readily listed in the loss column of the standard cost statement. For example:

Excavation to bases losing 7 h
Excavation to basement losing 7 h labour
Excavation to basement losing 3 h plant
Blinding losing 1 h
Make beam shutters losing 3 h

W/E 19/2/72										SHEET NO. 1	

WEEK NO. 3

M	T	W	T	F	S	S	TOTAL HOURS	MEASURE	@ h	STANDARD VALUE, h	DESCRIPTION	HOURS GAIN	HOURS LOSS
10	15		10	20			55	$12\ m^3$	4	48	Exc. bases by hand	—	7
18 / ⑨	18 / ⑨	18 / ⑨	10 / ④				64 / ㉛	$425\ m^3$	$^2/_{15}$ / $^1/_{15}$	57 / ㉘	Exc. basement by 22RB (plant)	— / ⊖	7 / ③
			8		15		23	$3\ m^3$	8	24	Concrete column bases	1	—
9							9	$2\ m^3$	4	8	Blinding	—	1
	7	18	21				46	$7\ m^2$	8	56	Fix shutters to columns	10	—
18	9		6				33	$6\ m^2$	5	30	Make beam shutters	—	3
				18	15		33	$28\ m^2$	1½	42	Strip base shutters	9	—
				9	8		17	$8\ m^2$	3	24	Make wall shutters	7	—
18	18	18	18	18	15		105	$45\ m^2$	2½	112	Brickwork to m/h. 1B thick	7	—

| | | TOTAL LABOUR | | | | | 385 | | | 401 | | 34 | 18 |
| | | TOTAL PLANT | | | | | ㉛ | | | ㉘ | | ⊖ | ③ |

Figure 4.1

Standard Costing

In addition to the individual performance of each operation by straightforward addition, it is possible to calculate from the standard cost what the contract's performance as a whole has been for the week being studied.

In the example, gains of 34 h and losses of 18 h have been made on labour, giving an over-all contract gain of 16 labour-hours for the week with an over-all loss of 3 plant-hours.

When the standard cost is calculated in man-hours, it is possible only to include those overheads that can be calculated in man-hours, e.g. travelling time, non-productive overtime, chainman, erection of temporary buildings, etc. For these a standard value can be calculated and a gain or loss shown. However, there are also numerous overheads that cannot be expressed in hours (except by converting pounds into equivalent hours), e.g. selective employment tax, lodging allowance, holidays with pay stamps, etc.

The means of calculating standards for overheads are dealt with in Chapter 7. *Figure 4.2* shows assumed standards illustrating a standard cost statement in nett cash for both measured work and overheads.

For practical purposes it is not necessary to quote cash figures to the nearest penny; gains or losses of less than £1·00 are relatively

W/E   19/2/72

WEEK NO. 3   SHEET NO. 1

M	T	W	T	F	S	S	TOTAL HOURS	@ 37½p = £ COST	MEASURE	@ £	STANDARD VALUE.£	DESCRIPTION	GAIN £	LOSS £
10	15		10	20			55	20.62	12 m³	1.50	18.00	Exc. bases by hand	—	2.62
18 ⑨	18 ⑨	18 ⑨	10 ④				64 (31)	24.00 @ 1.50 = 46.50	}425 m³	0.15	63.75	Exc. basement by 22 RB (plant)	—	6.75
		18	27	27	9		81	Carpenters @ 44p 35.64			20.00	Erect. temp. offices	—	15.64
from wages sheets								5.00			8.00	Tea woman	3.00	—
from wages sheets								24.00			16.00	Non-prod. Overtime	—	8.00
from wages sheets								45.50			31.50	Site staff	—	14.00

Figure 4.2

insignificant and can normally be ignored. Paperwork can thus be reduced if all extensions both of cost and of value are completed to the nearest pound. A further saving in computation can be made by averaging the cost per hour of tradesmen and labourers, in the ratio applicable to the contract. Within the limits of accuracy required this figure can then be rounded off to an easily calculable figure, e.g. 40p per hour, as illustrated in *Figure 4.3*.

M	T	W	T	F	S	S	TOTAL HOURS	@ 40p = £ COST	MEASURE	@ £	STANDARD VALUE, £	DESCRIPTION	GAIN £	LOSS £
10	15		10	20			55	22	12 m^3	1.50	18	Exc. bases by hand	–	4

Figure 4.3

It can be seen that this approximation causes a slight inaccuracy; but as the total expenditure on the standard cost sheet can at the end be balanced with the total expenditure on the wages sheets, etc., the amount of this inaccuracy can be readily seen and, if necessary, corrected. In practice the degree of control is rarely so fine that these approximations become critical.

In gross cash costing the gross labour rate is calculated at regular intervals, preferably weekly to tie in with the wages sheets, and will therefore reflect any shift in the labourer-to-tradesmen ratio. It is possible to calculate such a gross labour rate for different trades by separating the trades on the wages sheets. This may be necessary where, for instance, a particular trade is being paid exceptionally high bonuses, thus causing say a 10 per cent increase on the average rate for all other trades.

STANDARD COST STATEMENT

On many contracts the number of gangs and intermixing of items necessitates the rewriting of the numerous operations carried out each week under trade or section headings rather than gang headings—this provides an opportunity of reducing the cost statement to an easily readable form and thus encouraging site management to put it to use rather than to bed! *Figure 4.4* shows a typical standard cost statement rewritten under trade headings. Note the way information is kept to a bare minimum; this promotes action by encouraging the eye to run down the final column and

CONTRACT		W/E 29/1/72
SHEET NO. 1		WEEK NO. 28

COST CONTROL STATEMENT

DESCRIPTION OF WORK	VALUE £	COST £	GAIN £	LOSS £
Piling (Labour)				
Drive 300 mm x 300 mm piles	120	90	30	—
Drive 450 mm x 450 mm piles	30	32	—	2
	150	122	30	2
Piling (Plant)				
Drive 300 mm x 300 mm piles	90	80	10	—
Drive 450 mm x 450 mm piles	20	22	—	2
	110	102	10	2
Excavator (Labour)				
Excavate vaults	36	59	—	23
Excavate bases	12	50	—	38
Excavate trenches	15	14	1	—
	63	123	1	61
Excavator (Plant)				
Excavate vaults 22 RB	20	30	—	10
Excavate bases 22 RB	8	15	—	7
Excavate trenches JCB 4C	18	16	2	—
	46	61	2	17
Concretor (Labour)				
Blinding	80	60	20	—
Concrete bases	70	55	15	—
Drain bed and surround	20	15	5	—
Fence post holes	8	7	1	—
Columns	16	14	2	—
Beams	4	3	1	—
Ground floor slab	18	12	6	—
	216	166	50	—
Formwork (Labour)				
Bases	10	12	—	2
Bases (Credit)	50	—	50	—
Columns	10	8	2	—
Beams	11	7	4	—
Ground floor slab stop ends	15	13	2	—
	96	40	58	2
Formwork (Sub-let)				
Columns	30	15	15	—
Beams	42	25	17	—
Suspended slab	18	12	6	—
	90	52	38	—
Reinforcement (Sub-let)				
Cut and bend reinforcement	105	120	—	15
Fix bars	91	105	—	14
Lay fabric	15	10	5	—
	211	235	5	29
Bricklayer (Labour)				
Manholes	25	20	5	—
Scaffolding (Labour)				
Independent scaffolding	25	22	3	—
Scaffold ramps	4	4	—	—
	29	26	3	—

Figure 4.4

thus observe the losses. For additional effect the losses can be shown in red, or the losses column can be tinted red and the gains column tinted green. Such visual aids, although amusing at first, are bound to attract attention, which is all-important when a particular site manager is reluctant to use modern management tools or no follow-up system is built into the standard costing scheme as suggested in *Figure 15.1*, page 139.

CONTRACT		W/E	29/1/72
SHEET NO. 2		WEEK NO.	28
COST CONTROL STATEMENT			
DESCRIPTION OF WORK	VALUE £	COST £	GAIN £	LOSS £
---	---	---	---	---
Variable Overheads				
Non-productive overtime	32	68	—	36
Importation of labour	59	104	—	45
Inclement weather time	12	—	12	—
Holiday with pay stamps	34	30	4	—
Employer's National Insurance	23	20	3	—
National increases	26	—	26	—
Plus rates and extras	25	32	—	7
Graduated Pension	15	14	1	—
	226	268	46	88
Non-Variable Overheads				
Unload materials	20	68	—	48
Tea and canteen	5	20	—	15
Chainman	6	6	—	—
Plant maintenance	10	60	—	50
Pumping	5	14	—	9
Clean public roads	4	25	—	21
Clean office	2	12	—	10
Attend on Clerk of Works	2	12	—	10
Transport on site	10	50	—	40
Temporary lighting	3	—	3	—
	67	267	3	203
Plant Overheads				
Concrete mixer	6	8	—	2
Mortar pan	3	3	—	—
Coaches	12	22	—	10
Transport on site	10	40	—	30
Lighting set-up	1	—	1	—
Pumps	2	5	—	3
Vibrators	3	4	—	1
Power float	2	1	1	—
Compressor	3	2	1	—
	42	85	3	46

Figure 4.5

Overheads, both labour and plant, can be grouped together according to type and listed on a similar statement as illustrated in *Figure 4.5*. All statements can then be summarised and totals kept up to date of the various trades and overheads, as illustrated in *Figure 4.6*.

CONTRACT					W/E 29/1/72			
					WEEK NO. 28 OF 104			
	COST CONTROL SUMMARY SHEET							
	THIS WEEK				TO DATE			
	VALUE	COST	GAIN	LOSS	VALUE	COST	GAIN	LOSS
MEASURED WORK								
Piling (Labour)	150	122	30	2	2 024	1 740	320	36
(Plant)	(110)	(102)	(10)	(2)	(110)	(102)	(10)	(2)
Excavator (Labour)	63	123	1	61	1 206	1 080	220	94
(Plant)	(46)	(61)	(2)	(17)	(2 020)	(1 813)	(260)	(53)
Concretor	216	166	50	—	1 072	860	240	28
Reinforcement	—	—	—	—	860	630	370	140
Formwork	96	40	58	2	680	675	15	10
Bricklayer	25	20	5	—	700	650	50	
Drainage	—	—	—	—	250	190	70	10
Scaffolder	29	26	3	—	300	295	5	
Measured Bonus	—	—	—	—	—	—	—	—
Non-Productive Bonus	—	58	—	58	—	480	—	480
TOTAL MEASURED LABOUR	579	555	147	123	7 092	6 600	1 290	798
TOTAL MEASURED PLANT	(156)	(163)	(12)	(19)	(2 130)	(1 915)	(270)	(55)
TOTAL MEASURED	735	718	159	142	9 222	8 515	1 560	853
OVERHEADS								
FIXED LABOUR PRELIMS.	67	267	3	203	1 690	3 247	123	1 680
PLANT PRELIMS.	(42)	(85)	(3)	(46)	(700)	(986)	(29)	(315)
ON COSTS	226	268	46	88	3 075	2 878	758	561
TOTAL OVERHEADS	335	620	52	337	5 465	7 111	910	2 556
SUB-LET								
Reinforcement	211	235	5	29	450	505	55	110
Formwork	90	52	38	—	90	52	38	
TOTAL SUB-LET	301	287	43	29	540	557	93	110
GRAND TOTAL	1 371	1 625	254	508	15 227	16 183	2 563	3 511

GAIN/LOSS THIS WEEK £ 254 = 18½ %

GAIN/LOSS TO DATE £ 956 = 6¼ %

Figure 4.6

STANDARD COSTING IN CIVIL ENGINEERING

Where the plant element of a contract is likely to be extensive, e.g. in motorway and other large earthmoving contracts, the method of splitting the occasional items of plant illustrated in *Figure 4.4* becomes a little clumsy; two sets of figures continually have to be added together for each operation before a realistic comparison can be made. Labour and plant values and costs can be added together for such a statement, although provision will have to be made elsewhere for the following factors.

1. In nett costing the total labour value is required for calculating the variable overheads. This can easily be abstracted from the standard cost statement when labour and plant are separated but will have to be noted separately if they are combined. This can be done in the margin or in an additional column. Alternatively, the variable overheads can be related to the labour *and plant* element of the contract. However, this is a somewhat false relationship and would show unrealistic results on a contract having a large plant element; for example, extensive earthmoving by scraper would increase standard value for selective employment tax (S.E.T.) stamps, etc., because of increased plant standard value in measured work. Thus control of variable overheads would not be possible, as they would not be directly related to pure labour values.
2. Balance of labour costs with wage sheets would not be directly possible. However, if the cost of plant on site is recorded separately, a balance can be made between total costs in the standard cost statement and wages sheets plus plant costs plus cost of labour-only sub-contractors.

Exercises

1. Discuss the costing systems known to you and decide whether they are standard costing or unit costing or a mixture of the two systems.

2. State which of the two systems, standard costing or unit costing, you would prefer to use as a management tool and give details why.

5
EXPENDITURE DATA FOR UNIT COSTING AND STANDARD COSTING

The three elements of expenditure described in Chapter 1 as requiring regular control are:

1. labour costs;
2. plant charges;
3. site overheads.

A breakdown of expenditure for these three elements is required before either a unit costing or standard costing system can commence; hours of plant and labour require allocation against individual operations of work. There are a number of ways of obtaining this allocation.

LABOUR COSTS

Task Sheets

Daily task sheets may be used for each man or gang of men. *Figure 5.1* illustrates a task sheet designed for issue to each man as he clocks on in the morning.

As he starts or finishes each operation, the worker notes down the approximate time, say to the nearest 15 or 30 min; alternatively he notes down the duration of each operation. Should his programme be altered for some reason, he can easily add additional items to the programmed list as in the example: the 're-erection of the signboard accidentally knocked over by the contractor's plant'.

Where a gang of men are all carrying out the same operations with only occasional variation by members of the gang, then this sheet can be satisfactorily used for that gang with any variations noted on the sheet.

Expenditure Data for Unit Costing and Standard Costing

ANY FIRM & SON LTD.		CONTRACT	
DAILY TASK SHEET			
PROGRAMME FOR DAY		START	FINISH
Remove door from Superintendent's office	D/W	9.00	9.30
Rehang door when safe installed	D/W	3.00	4.00
Fix skirtings to Rooms 23 and 24		8.00 9.30	9.00 11.00
Hang doors A13 and A14		11.00 1.00	12.30 3.00
Fix pipe casing in NW corner of Room 11		–	–
Re-erect signboard knocked down by J.C.B.		4.00	5.00
NAME: D. Peters		DATE: 4/12/71	

Figure 5.1

For example, in an excavation gang of six men two of the gang stay behind in a base excavation to bottom up while the others move to their next job of work. This can be shown as in *Figure 5.2*.

PROGRAMME FOR DAY	START	FINISH
Excavate base 3C	8.00	10:30 (4) 11.30 (2)
Excavate drain trench	10.30 (4) 11.30 (2)	12.00

Figure 5.2

If durations only are shown, the format can be that of *Figure 5.3*.

PROGRAMME FOR DAY	NO. OF MEN	DURATION
Excavate base 3C	4	2½
	2	3½
Excavate drain trench	4	1½
	2	½

Figure 5.3

Expenditure Data for Unit Costing and Standard Costing

The larger the gang and the more diverse the operations, the more difficult such a simple task sheet becomes to control. A pour of concrete, for instance, may have a large number of attendance items required before, during and after the main function of placing the concrete: for example, preparation of construction joints; travelling crane from last concrete pour; preparation of barrow runs for area out of crane's reach; preparation of vibrators; pouring and vibrating concrete; tamping concrete; clearing up surplus concrete and barrow runs; trowelling; rubbing up; removing wet sand used for curing.

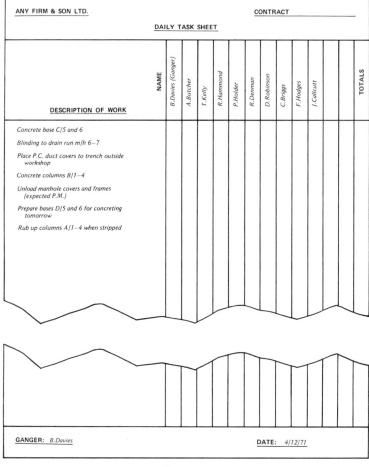

Figure 5.4

For this reason, and to obtain greater accuracy, it is better to show the duration of *each* man on *each* operation.

Figure 5.4 illustrates a gang task sheet as written up by the planning engineer or similar person, and based on the site's over-all programmes and resources.

Discussion of the problems of daily site planning is outside the scope of this book, but it is, however, necessary to point out the uselessness of producing daily task sheets based solely on programme requirements without first discussing the work with the foreman concerned, whose intimate knowledge of available resources, short-cuts, probable absenteeism, individual's capabilities, etc., can turn a random list of operations into an enthusiastically attempted programme of work.

Where a job card system (page 64) is in operation for incentive purposes times can be entered on the job card either against a total job or allocated against the various tasks within that job as, for example: shutter walls—stop ends, surface fixings, height lath and joint key.

Ganger's Allocation Sheets

Where site planning or team co-operation have not yet reached the level necessary to produce daily task sheets for a contract, and therefore individual operations cannot be pre-listed, an attempt must be made to classify work under the required descriptions after that work has been carried out. This can be done by relying on the ganger to write down the descriptions of what his men have been doing.

Figure 5.5 illustrates a ganger's daily allocation sheet similar to the daily task sheet but completely written out by the ganger or foreman. As with the gang task sheets, a column is allowed for the time spent by each man on each operation. The reduction of columns to a single gang column, although reducing the size of the form, would undoubtedly either create confusion by the ganger attempting to reconcile, for instance, 1 man @ 3 h, 2 men @ 6 h, 2 men @ 4 h, and 1 man @ 2 h; concreting the basement slab, or, more likely, lead to approximations that might end up by the ganger roughly spreading the man-hours among the operations, disregarding all limits of accuracy.

When written out as suggested by site men, the legibility and phraseology of these sheets can, not surprisingly, be something of a problem, though not without its lighter side. Comments of 'lifting off Irish Jays' and 'burning out Polish Irene' are perhaps understandable, although the recently seen entry of 'sissy engineers'

Expenditure Data for Unit Costing and Standard Costing

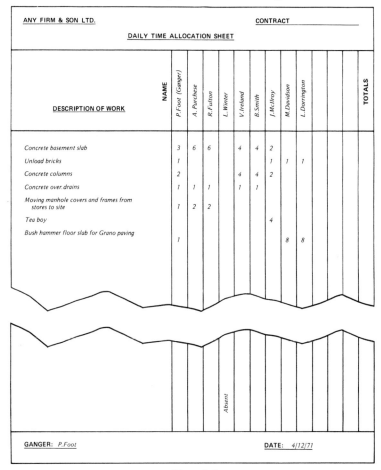

Figure 5.5

when the chainboy was sick and a labourer came to the engineer's *assistance* makes one wonder whether the degree of illiteracy is not intentional!

By using (*a*) standard descriptions or (*b*) coded descriptions, it is possible to alleviate this problem to some extent, though at the same time creating others.

Standard descriptions are in a way an extension of the daily task sheet; but instead of listing only those operations programmed for the day, the sheet lists every possible operation that could be carried out on the contract. For reasons of pure length the lists have to be split into trades or some other division, and even then can amount to some half-dozen sheets of possibilities for a single trade. With carpenters, for example, make, fix and strip, and every different type of shutter are listed separately. With excavators dig, plank and strut, trim and backfill are listed separately. This, of course, is true only of a large contract, but then a small contract hardly requires the use of standard descriptions. As with the daily task sheets, space must be left at the end for the ganger to write in unlisted work. *Figure 5.6* illustrates a standard description allocation sheet for the steelfixer trade on a particular contract.

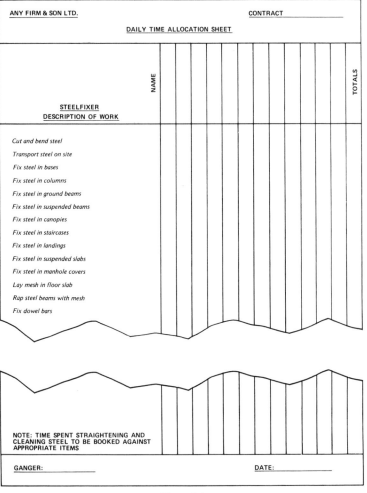

Figure 5.6

Expenditure Data for Unit Costing and Standard Costing

Coded descriptions remove the problem of incomprehensible allocation sheets by allocating a code to every operation; thus the ganger or foreman only has to write down the code and will not be tempted to ramble on into illegibility.

A possible code might start with the identification of the trade:

E. — Excavation	R. — Reinforcement
C. — Concrete	D. — Drains
F. — Formwork	B. — Brickwork
	etc.

The next two letters of the code would show the location:

ab. — Abutment	dr. — Drain
bs. — Base	sl. — Slab
bt. — Basement	wl. — Wall
bm. — Beam	rf. — Roof
cl. — Column	etc.

For example, C.cl. means concreting to columns; C.bt.rf. means concreting to basement roof.

For the majority of items this kind of code is sufficient; and provided that details are placed in a prominent position on the wall of the ganger's or foreman's office, it is quite possible to operate such a scheme. However, a lot depends on the temperament and education of the ganger or foreman, and it is suggested that an engineer or timekeeper assist in the filling in of such an allocation sheet.

An alternative suggestion is a numerical code:

1. Excavation
2. Concrete
3. Formwork
4. Reinforcement
5. Drainage
6. Brickwork
etc.

.01 Oversite
.02 Reduce levels
.03 Basement
.04 Bases
.05 Footings

with further figures giving dimensional or other further details.

For example, 1.02 means excavate to reduce levels; 4.04 means reinforcement to bases.

A numerical code as suggested above is usually not capable of being used at foreman level, allocation being made in the normal long-hand way and coding being added at office level later for handling by computer.

In an attempt to operate the ganger's allocation system speedily and accurately, a ganger is frequently given some additional incentive, e.g. an additional $\frac{1}{4}$ hour's pay per day or an additional 1p/h on his rate. It is important that the conditions under which the ganger or foreman receives this payment be fully known and enforced. For example: Payment only made if allocation sheet is on agent's desk by 10.00 a.m. on day following work being allocated. No sheet—no pay, late sheet—half pay.

Where bonus schemes are in operation the tendency of the ganger may be to book hours against non-bonusable operations; for this reason an attempt must be made to bonus all work so that the ganger will realise that nothing can be gained from this practice.

Perusal of the allocation by the ganger's or foreman's superior will discourage inaccurate or sloppy allocation; and as a check on the accuracy of the system, a spot cost on an individual gang during the course of a day will prove sufficient provided that the foreman is not aware that the check is being carried out.

Personal Observation

Personal observation of the distribution of men is made by one of the following three basic methods.

1. Instead of the ganger or foreman filling in an allocation form, the responsibility is taken over by a timekeeper, engineer or surveyor, and it is then his duty to patrol the site at frequent intervals in order to note down any changes in labour and plant activities. This system is necessary where site personnel are illiterate or are reluctant to become involved in paperwork. It also alleviates the problem of unrecognisable descriptions, and enables the timekeeper, engineer or surveyor to become familiar with the workings of the site. It is clear that on even a medium-sized contract such personal observation could become a full-time operation, and on an extensive development or civil engineering contract could require such a large staff that the method would be uneconomical.
2. Staff requirements can be reduced by requiring the time-

keeper, engineer or surveyor to patrol the site less frequently, say once during the morning and once during the afternoon. The frequency is now too long to be able to record sufficiently accurate movements of labour. Some operations may be completely carried out between observations and would therefore be missed. However, the personal observations can be supplemented by verbal information from the ganger or foreman. A five-minute chat between, say, the timekeeper and the ganger is a lot less expensive than an hourly trip round the site by the timekeeper.
3. On some civil engineering sites allocation of resources may be unnecessary, as a gang may be working on the same operation for a number of weeks or even months. Observations are then necessary only for method study or for incentive bonus payments. Costing can therefore be carried out directly from the timebook.

PLANT CHARGES

Except for special studies the majority of small plant items on a site do not need to be allocated hourly. Little is to be gained by allocating, for instance, a poker vibrator to each individual pour of concrete or a pump to the various items of excavation. These can be charged to the contract on a weekly basis and costed as plant overheads. Large plant, on the other hand, can be allocated on an hourly basis in the same manner as labour, as illustrated in *Figure 5.7*, overleaf.

Mixers, tower cranes, etc., can be similarly allocated, though each contract and each item of plant poses the question: 'Is the information to be gained from such allocation worth the effort involved?' It is not necessary during a regular costing system to know how long a mixer is working for a certain pour of concrete, nor what a tower crane is lifting during each hour of the day. This kind of study is best left as a method study exercise and not incorporated in the costing system. It is suggested therefore that allocation of plant be generally confined to excavation and piling equipment only.

The rates used for calculating the cost of plant, whether allocated or weekly, must be the true cost to the contract. If the plant is hired, then the hire charge applies; if the plant is owned by the contractor, then the contractor's internal hire rate applies. It is not normally necessary to approximate between these two sets of charges unless, for instance, two or more identical machines are being used on the site, one hired and one contractor-owned. The

ANY FIRM & SON LTD.				CONTRACT						
DAILY TIME ALLOCATION SHEET										
NAME	P. Holder (Dr)	L. Hennessy (B/M)	22 RB							TOTALS
Excavate basement	4	4	4							
Excavate lift base	5	5	5							

Figure 5.7

use of two separate rates may cause unfair weighting against one operation in favour of another. For example, two of the contractor's own excavators are digging bases but have lost approximately two weeks of programme. Duct excavation is due to start and cannot be delayed until one of the excavators becomes free in two weeks' time. The contractor has no further machine available; therefore a machine is hired for two weeks and put to work excavating the ducts.

It would be a true cost to charge the higher rate against the duct excavation but a fairer result might be obtained by averaging out the excavator rates during the two-week period that the hired machine is on site.

It is sometimes beneficial to show such an 'Increase in plant costs' under the plant overheads section of the cost statement, thus highlighting the cost of deviations from the contract programme.

Another item of expenditure which can be included under the heading of 'Plant overheads' is the cost of fuel, oils, etc. This is preferable to adding an estimated cost of fuels, etc., to the plant hire charge rate, because

1. the plant rate then becomes partly fact and partly estimated, and

2. any losses of fuel costs would be concealed by the plant costs.

A difficulty arises when plant is standing owing to lack of work, to breakdown or to lack of driver. Should the means of hiring or the contractor's internal hiring policy necessitate a charge to the contract during standing or broken-down time, then this charge is best shown as a plant overhead and possibly labour overhead of 'Standing time'. It is unrealistic to levy such charges against any particular operation that happened to be under way at that time or, worse still, happened to be showing a good output at that time. Note should be taken here of any differentials applicable to plant charges when the plant is not working or of any minimum working time agreements between the contractor and the plant hire-company or department.

SITE OVERHEADS

Site overheads fall into three basic categories:

1. wage sheet overheads;
2. allocated labour overheads;
3. plant overheads.

Wage Sheet Overheads

NETT COSTING

For use in standard cost statements the wage sheet overheads can readily be abstracted from the wages sheets: for example, total cost for week of S.E.T., lodging allowance, sickness payments, holiday stamps, etc.

Flat rate overtime working must be allocated in the same manner as normal working hours, but the premium time (excess overtime or non-productive overtime) requires to be kept separate and calculated as an individual cost item.

Wage sheet overheads cannot be included in nett unit costing except as an addendum list at the end of the unit cost statement.

GROSS COSTING

Where the unit or standard cost is to be produced 'gross' as described in Chapters 3 and 4, then wage sheet overheads become part of the average gross rate and are not studied individually.

Allocated Labour Overheads

NETT COSTING

Whatever system is used for allocating labour costs hour by hour, certain non-measurable items will be disclosed which are best dealt with as an overhead cost. These items include erection of site office, unload materials, lay temporary roads, clean public roads, etc. In unit costing these items can only be listed as an addendum to the unit cost statement; however, in standard costing it is possible to incorporate such items into the general standard cost statement under the heading of 'Overheads'.

GROSS COSTING

In both unit costing and standard costing these items can be listed as an addendum to the gross cost statement. Alternatively, when the average hourly rate is calculated the wage sheets can be divided by the total of the productive work hours only, i.e. total hours less allocated labour overheads; the gross rate will then be slightly higher and will include these overheads.

Plant Overheads

NETT COSTING

Reference has already been made to suggested items of plant overheads (page 35). Such items as vibrators, fuel and oils, mixers, dumpers, etc., can be abstracted from the site plant book, records of plant on site or invoices without need for allocation. Plant standing or broken down can be recorded in the same way that the plant working is allocated. It is not usual to record the cost of vibrators, dumpers, etc., standing as a separate item, as these are already an overhead and are probably charged anyway on a weekly basis.

In unit costing these items, too, can only be listed as an addendum to the unit cost statement; however, in standard costing it is possible to list such items under a heading of 'Plant overheads'.

GROSS COSTING

Plant overheads are usually dealt with as described under nett costing, though as an extension of the gross labour rate an addition can be made to cover overhead plant charges, i.e. the calculation of the gross labour rate:

$$\text{g.l. rate} = \frac{\text{wage sheet total} + \text{overhead plant charges}}{\text{number of productive hours worked}}$$

Expenditure Data for Unit Costing and Standard Costing

The result can be quite a high average gross rate out of all proportion to the basic labour rate but nevertheless including all overhead costs, both labour and plant.

Exercises

1. Design a system of collecting labour and plant allocation for use on each of the following sites. What instructions would you issue for ensuring correct working of the system?

£120 000	Block of flats
£20 000	6 No. detached houses
£7 000 000	10-mile motorway
£500 000	Sewage works
£30 000	Car factory maintenance
£12 000 000	Reinforced concrete dam

2. Bulk earthmoving requires special attention due to the high plant costs and therefore critical need for high outputs. Draw up a form for collecting the allocation of earthmoving plant by personal observations (load checking), bearing in mind that the different types of excavation—topsoil, suitable clay, unsuitable ripped mudstone, etc.—and different haul lengths and condition of haul route require to be recorded by the checker if any worthwhile feed-back of information is to be gained.

3. Design a job card suitable for recording detailed allocation of tasks within that job and for indicating which men have carried out the job.

6
STANDARDS FOR MEASURABLE WORK

The previous chapter has shown the source of the various elements of cost for use in both unit costing and standard costing.

However, the basis of a standard costing system is the comparison of these costs with some yardstick or standard, and the question immediately arises: 'With what can the costs be compared?'

LIBRARY OF HISTORICAL OUTPUTS

Under factory conditions it is possible to record consistency of outputs which will never be achieved on a construction site. However, this lack of consistency is not a reason for ignoring historical records but more a reason for conserving as much data as possible under all types of conditions on all types of sites, the relevant conditions being recorded with the output as indicated in *Figure 6.1*.

Extreme differences of conditions are worthy of separate recording; for instance, it is pointless to list together excavate bases in sandstone and excavate bases in granite. The average of these two items would be a meaningless figure. The cards or forms containing similar types of items, as in this example, can, however, be filed next to one another, and a handy record of outputs for similar items under differing conditions will thus be preserved.

It is important to indicate any special points about a contract which may have affected the average output recorded—for example, exceptional weather conditions, abnormal requirements of architect or engineer, use of explosives, errors not possible to separate from output, use of new methods, etc. Even the name of the foreman in charge may be found to have a consistent effect on the output. It is important, too, that the calculation of the output be over as

OPERATION			J.C.B. EXCAVATION TO BASES AND PITS IN CONSOLIDATED FILL							
DATE		CONTRACT	DESCRIPTION OF CONTRACT	SPECIAL REMARKS	QUANTITY	LAB. HOURS	J.C.B. HOURS	HOURS	HOURS	OUTPUT
FROM	TO									
Jan. 69	Mar. 69	Office Block	R.C. frame £200 000	Industrial area. Approx. 25% broken brick in fill Aver. base size 6 m^2 × 3 m deep	450 m^3	135	40			11¼ m^3/h
Feb. 70	June 70	Factory	Steel frame £120 000	Unsuitable fill from motorway 75% clay Aver. base size 4 m^2 × 3 m deep	380 m^3	120	35			11 m^3/h
July. 71	Dec. 71	Bowling Alley and Skating Rink	R.C. frame £300 000	Industrial area Approx. 10% ash in fill. Aver. base size 9 m^2 × 4 m deep	800 m^3	170	65			12½ m^3/h

Figure 6.1

long a period as possible in order to avoid the recording of brief abnormalities. The ideal period for recording output data is no less than from start to finish of the operation being studied. This reveals the only true average output: at the middle of the operation everything is in full swing; at the beginning the operation is having teething troubles; at the end of the operation cleaning up and finishing off is being carried out.

WORK STUDY STANDARDS

Where work study engineers are employed or a work study data bank is available, an ideal situation exists for comparing actual costs with a scientifically calculated standard. It must, however, be emphasised that work study standards derived from one contract must be used on other contracts with extreme caution, methods must be fully defined in all records and a consistent approach given to relaxation allowances, supervisory time, travelling time, labourers in attendance and unoccupied time which occurs during team work when a man or machine is forced to be idle even though the man: machine ratio is at the optimum.

Figure 6.7, page 50, shows the clauses of the British Standard relevant to time study techniques of work measurement. For use in standard costing either the 'allowed time' or 'standard time' can be used, provided the percentage difference is clearly understood, standard time being based on a motivated rate of working (B.S. 3138:1969 Clause 34016) and allowed time (B.S. 3138:1969 Clause A1017) being an easing of that standard usually for payment or cost control purposes. On a poorly organised site, a site with little incentive or low team morale, few actual outputs are likely to reach the work study standard, thus making allowed time a more realistic yardstick for control.

ESTIMATOR'S PRICING

Where no adequate record of outputs is available either from past costing records or from work study, an alternative basis must be found for comparison with costs, the most likely alternative being the data used in pricing the contract tender.

Consideration must first be given to whether the standard cost based on estimator's rates is to be run

Standards for Measurable Work

1. as a *pure standard cost,* i.e. by simply using the estimator's standards as a guide and varying the standards according to conditions as found on the site, or
2. as a strict *profit and loss account* adhering rigidly to the prices and outputs used in the tender, varying only on the issue of a variation order by the client.

For the majority of operations these two methods are identical; but without the historical data necessary to review standards scientifically the frequent varying of standards suggested in (1) above can be abused and can lead the costing system into fantasies that are not respected by site management. It is preferable therefore, when estimator's standards are used, to adhere strictly to known figures whatever the circumstances and whatever the changes on site, as suggested in (2) above.

In breaking down the estimator's rates, care must be taken to return all adjustments to their appropriate places ready for use in the costing system under the correct heading.

Various adjustments may have been made to the estimator's original calculations; these generally fall under the following headings.

ADDITION OF PRELIMINARIES TO THE RATES

For example, the initial cost of setting up offices, mixers, site roads, etc., has been added to the excavation items in order to recover the cost early in the contract; before excavation rates are used as costing standards, the allowance for these items must be deducted, to leave only the excavation element of the rates.

INTERNAL COMPANY ADJUSTMENTS

For example, the pricing may have been adjusted to allow for using written-off plant; the plant standard must therefore be nil or whatever nominal amount has been left in the pricing. If other plant has to be used, e.g. hired plant, then a loss will correctly show up against the standard for the operation using that plant.

FEEL OF THE MARKET

Because of a full order book, because of the need to acquire work to give continuity, because of goodwill, or because of the feeling that other contractors are fighting for a contract or are not interested in a contract, a firm may well adjust a series of rates or the final figure of their quotation or tender. If a true standard cost system is to be operated, this adjustment must not affect the setting of the standards; however, if a profit and loss standard cost system is to be operated, it is preferable to make this adjustment where appropriate, thus observing the strict rule of no variations on known figures. It can be argued, however, that such an adjustment as this cuts into or enhances the planned profit margin and should therefore not affect anticipated outputs. This is, of course, true but increased or decreased outputs have nevertheless got to be achieved to make up for the adjusted profit margin.

CLIENT'S EFFECT

An adjustment may be made within a quotation or tender because of the anticipated ease or difficulty of working with a certain client, architect or consultant. Such an adjustment must also be made to the costing standards in anticipation of higher or lower outputs during construction.

ERROR

Errors in pricing usually have to be accepted by the contractor and in a profit and loss standard cost must be carried through to the standard. In a pure standard cost, however, the error can be corrected and the standard can be made realistic. With the introduction of the metric system the old bogey of pricing yard super items with a foot super price has, of course, disappeared, but care must still be taken to see that pounds have not become pence or thicknesses of concrete, timber, brickwork, etc., misread.

LATE INFORMATION

Although last-minute quotations from sub-contractors and suppliers do not usually affect the labour, plant and overheads section of the pricing, a late labour only sub-contractor may well submit a price in time for an adjustment to be made on the tender figure. Such an adjustment must be carried through to the costing standards for sub-let work.

Standards for Measurable Work

DEVIATION FROM NORMAL

For example, excavation rates may as a matter of course include an allowance for pumping. If this is removed or increased on the tender figure because the site is exceptionally dry or exceptionally wet, then this adjustment must also be made to the standard for pumping.

WEIGHTING OF MONEY

For financial reasons a contractor may reduce the rates for certain items, transferring the money to other items in the bill of quantities or quotation. This money must be returned to its rightful place before standards are calculated. Even where the standard cost is of the profit and loss account type, this deviation from the billed prices back to estimator's prices is desirable, as curious gains and losses will obviously result from the use of weighted rates or outputs. Any increases or decreases on billed quantities of weighted items can be studied outside or as an addendum to the normal standard cost system, e.g. abstracted from the monthly valuations.

Breakdown of Estimator's Rates

Methods of estimating vary tremendously throughout the industry from rates worked out on the back of a cigarette packet to rates calculated entirely 'by computer'. Rates may be based on a hunch or on work study; but however they have been arrived at, a breakdown must be carried out to discover the standards that will subsequently be used in the standard costing system. Two elements require extracting from the bill or quotation rate—labour and plant—either gross or nett, depending on the proposed standard cost.

Because a standard cost is an attempt to compare like with like, the planned profit margin allowed in the estimate must be removed if standards are being calculated in cash. The same applies to the allowance for Head Office overheads, unless these overheads are being included on the cost side of the standard cost, e.g. added into the calculation for the gross labour rate. The method of estimating will dictate where and when this deduction takes place.

Figure 6.2 illustrates an estimator's pricing of an item for hardcore fill to a lift-shaft and is based on the gross labour rate shown in Example 7.1 on page 54.

Standards for Measurable Work

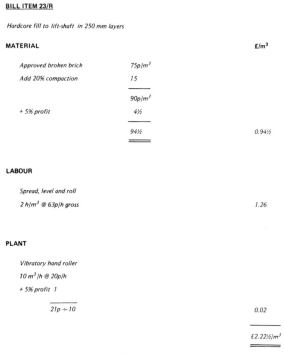

BILL ITEM 23/R

Hardcore fill to lift-shaft in 250 mm layers

			£/m³
MATERIAL			
	Approved broken brick	75p/m³	
	Add 20% compaction	15	
		90p/m³	
	+ 5% profit	4½	
		94½	0.94½
LABOUR			
	Spread, level and roll		
	2 h/m³ @ 63p/h gross		1.26
PLANT			
	Vibratory hand roller		
	10 m³/h @ 20p/h		
	+ 5% profit 1		
	21p ÷ 10		0.02
			£2.22½/m³

Figure 6.2

Breakdown for gross standard costing for this item is as shown in *Figure 6.3*.

Gross labour standard is 2 h × 57p, i.e. labour rate before addition of profit and head office (H.O.) charges (see page 55).

Gross plant is $\frac{1}{10}$ × 20p. No overheads have been added to plant in this estimate.

BILL ITEM	BILL RATE, £	GROSS LABOUR STANDARD, £	GROSS PLANT STANDARD, £	OTHER ELEMENTS INCL. ALL PROFIT AND H.O. CHARGES, £
23/R	2.22½	1.14	0.02	1.06½

Figure 6.3

Standards for Measurable Work

Other elements are the addition of:

	£
Material	0·94½
Profit on labour and H.O. charges—2 h @ 6p	0·12
Profit on plant—$\frac{1}{10} \times$ 1p	—
	£1·06½

The final column is provided as a balancing column to check mathematical errors.

Breakdown for nett standard costing for the same item is as shown in *Figure 6.4*.

BILL ITEM	BILL RATE, £	NETT LABOUR STANDARD, £	NETT PLANT STANDARD, £	OTHER ELEMENTS INCL. ALL PROFIT OVERHEADS AND H.O. CHARGES, £
23/R	2.22½	0.75	0.02	1.45½

Figure 6.4

Nett labour standard is 2 h × 37½p, i.e. basic labour rate.
Nett plant is $\frac{1}{10} \times$ 20p.
Other elements are the addition of:

	£
Material	0·94½
Overheads, profit on labour and H.O. charges 2 h @ 25½p	0·51
Profit on plant—$\frac{1}{10} \times$ 1p	—
	£1·45½

Breakdown by hours instead of cash for the same item is as shown in *Figure 6.5*.

BILL ITEM	LABOUR STANDARD h/unit	PLANT STANDARD units/h	TYPE OF PLANT
23/R	2 h/m³	10 m³/h	Vib. roller 20 p/h

Figure 6.5

Standards for MeasurableWork

By breaking down each item to be costed in this way it is possible to compile a complete list of labour and plant standards for every measurable operation envisaged during the course of the contract.

TEXTBOOK STANDARDS

Where neither an adequate library of standards nor estimator's calculations are available, e.g. in spec. building, package deal or where a lump sum contract has been negotiated, then the only remaining source of standards is from one of the many textbooks on estimating backed by individual's records and knowledge of outputs.

This system of setting standards, although approximate, does provide a degree of control over the site expenditure in that costs are compared with a constant and variations from that constant are easily seen and can be further studied.

Exercise

Break down the following five examples of bill pricing under the headings of labour and plant (*a*) nett, and (*b*) gross, where 10% of the addition for profit and overheads accounts for profit and Head Office overheads.

Example 6.1. Estimated from first principles
25 mm rough boarding nailed to rafters

Material

	£
Rough boarding	0·55
Waste—7½%	0·04
Nails	0·01
	0·60
Profit—5%	0·03
per m²	£0·63

Labour

Unload and fix	£
Carpenter 0·30 h @ 44p	0·13
Labourer 0·15 h @ 37½p	0·05½
	0·18½
Profit and overheads—100%	0·18½
Material	0·63
per m²	£1·00

Standards for Measurable Work

Example 6.2. Estimated from sub-let prices
1:2:4 concrete in bases

	£
Material (incl. 5% profit)	4·62
Labour and Plant	
Mix and place concrete sub-let	1·25
Add profit and overheads—20%	0·25
per m³	£6·12

Example 6.3. Estimated from labour constants
215 mm brickwork in 1:3 cement mortar

Material

	£
Bricks $\dfrac{£10\cdot00}{1\,000} \times \dfrac{125}{1}$	1·25
Mortar 0·1 m³/m² (1B) @ £5·00 per m³	0·50
Material	1·75
+ 5% profit	0·09
	£1·84

Labour

	£	£
Standard £12·00 per thousand	12·00	
Add 10% for difficulty on this contract	1·20	
	£13·20	
$£13\cdot20 \times \dfrac{116}{1\,000}$		1·53
Profit and overheads—100%		1·53
		3·06
Material		1·84
per m²		£4·90

Example 6.4. Estimated nett with profit and overheads added at end of bill of quantities

BILL REF.	QUANTITY	LABOUR	PLANT	MATERIAL
16/A	250 m²	0.90	0.15	3.75
B	35 m²	0.95	0.15	3.75
		+ 100%	+ 5%	+ 5%

Figure 6.6

Example 6.5. Estimated off the cuff

Excavate interceptor pit   £1·25 per m³

Group Studies

1. Within the definitions given in B.S. 3138 : 1969 *Glossary of Terms Used in Work Study* compare the various systems of adding allowances to basic times which are in use by companies and advisory services in the construction industry (see *Figure 6.7*).

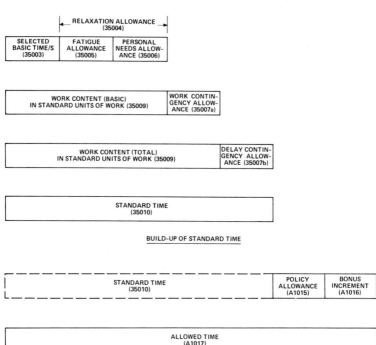

Figure 6.7. Figures in parentheses refer to clause in B.S. 3138 : 1969 (reproduced by permission of the British Standards Institution, 2 Park St., London W1A 2BS, from whom copies of the complete standard may be obtained)

2. Discuss the various methods of estimating known to you and the problems arising when these estimates are later used as a basis

Standards for Measurable Work

for standard costing. Decide on the method most suitable for both build-up and breakdown, bearing in mind that not every tender is successful.

Exercises

1. Assuming that no output library is available and that a factory contract has been placed on a price per square metre, without detailed estimating, use any available textbooks, etc., to research and calculate standard outputs in h/m^2, for use in a standard costing system for the following items of shuttering. Quantities have been roughly measured. Calculate, make, fix and strip separately:

300 m^2 ground beams
500 m^2 pile caps
100 m^2 edge of 200 mm floor slab
100 m^2 construction joint in 200 mm floor slab
350 m^2 walls
250 m^2 columns
300 m^2 suspended beams
100 m^2 suspended slabs
 10 m^2 suspended landings

2. Show how your approach to standards would differ when using estimator's prices as the basis of a true standard cost system from using the system as a profit and loss account.

7

STANDARDS FOR SITE OVERHEADS

Although historical data on site overheads may prove useful in estimating such items as importation of labour, attraction money, sick pay, etc., in certain regions or for certain types of contract, the majority of overheads will be peculiar to each contract and will usually be calculated from first principles.

The amounts allowed in an estimate for overheads are therefore the most reliable means of calculating standards for overheads. In the absence of such an estimate, budgeted figures must be used based on the same reasoning that an estimator would have used had he been pricing the contract in detail.

Site overheads can be divided into two distinct types: (*a*) fixed and (*b*) variable.

FIXED OVERHEADS

Few overheads are rigidly fixed during the full period of a contract but many remain more or less unchanged by the current labour strength at any particular time in the contract. These are termed fixed overheads, and include items of both labour and plant. Examples are:

Site offices	Chainboy
Toilets	Attendance on clerk of works
Temporary roads	Clean public roads
Staff	Plant maintenance
Minibuses	Temporary services
Tea woman	Small tools

These items are frequently priced under the preliminaries section of the bill of quantities and are therefore often referred to as 'Preliminaries' in a standard costing system.

Standards for Site Overheads

The estimated cost or budget for these overheads can be spread over the contract to obtain weekly standards. The following examples show how that budget can be spread.

1. Weekly allowance, e.g. cleaner for offices.
   £5·00 per week allowed for duration of contract.
2. Limited weekly allowance, e.g. chainboy.
   £6·00 per week for first 20 weeks of contract.
3. Lump sum allowance, e.g. initial site survey.
   £150·00 allowed for survey.
4. Measurable allowance, e.g. scaffolding.
   10p allowed for scaffolding per square metre of brickwork constructed.
5. Random allowance, e.g. temporary roads.
   £300·00 total allowed to be drawn on at random.
6. Allowance related to value of work, e.g. minibuses.
   3% on nett labour value of work completed.
7. Combination of above, e.g. temporary offices.
   £250·00 erect offices,
   £100·00 dismantle offices,
   £5·00 per week hire for duration of contract.

VARIABLE OVERHEADS

A number of overheads are geared directly to the number of men employed on site and vary as that number changes. These are termed variable overheads and include such items as:

Annual holiday stamps	General insurances
Public holiday stamps	Selective employment tax
State pension	Sick pay
Inclement weather time (I.W.T.)	Redundancy pay
Travelling time	National Insurance
Fares	National increases
Lodging allowance	Training levy

Some of these items, e.g. general insurances and training levy, may not be dealt with at site level and would therefore become part of the Head Office charges not included in the standard cost.

The pricing of these variable overheads is frequently calculated as an on-cost to the basic labour rates and the bill of quantity items priced gross, i.e. including the appropriate share of these overheads as in the pricing of 'Hardcore to lift-shaft' (*Figure 6.2*), which is based on the labour rate calculated in Example 7.1. Variable overheads are therefore often referred to as 'On-Costs' in a standard costing system.

In Example 7.1 for every 37½p worth of labourer work carried out the contract will earn an additional ½p to cover the cost of boot money, etc., an additional 9·8p to cover the cost of stamps, and other additional sums to cover other variable overheads.

The ½p can be expressed as 1·34% of 37½p, which provides an accurate relationship between value or standard of an individual overhead and the value or standard of measured work completed for the current week or period being studied.

The percentages of 37½p for the other variable overheads shown in Example 7.1 are as follows:

Attraction money	Nil
Holiday stamps	5·87%
National Insurance	4·80%
Graduated pension	2·66%
Selective employment tax	12·80%
Non-productive overtime	10·00%
Importation	12·00%
Wet time	2·98%
Redundancy	0·27%

Example 7.1

Labourer's rate			£
Labourer's basic rate			0·3750
Plus rates (boot, wet, cement, etc.)			0·0050
Attraction money (policy bonus)			—
Stamps		£	£
Holiday stamps, public		0·40	
annual		0·70	
National Insurance		0·90	
Graduated pension	ave.	0·50	
S.E.T.		2·40	
Insurances and training levy (if paid for by site)		—	
		£4·90	
Assume 50 h week $= \dfrac{4·90}{50}$			0·0980
Non-productive O/T $\dfrac{10 \times \frac{1}{2} \times 37\frac{1}{2}\text{p}}{50}$			0·0375
Importation			
Assume all men 1 h day travelling 6 day week			
$= \dfrac{6\text{ h @ }37\frac{1}{2}\text{p}}{50}$			0·0450
			£0·5605

Standards for Site Overheads

Wet time

Allow for guaranteed 40 h week and I.W.T. 2%			0·0112
Redundancy—allowance ¼%			
basic	0·3750		
non-productive O/T	0·0375		
	0·4125 × ¼%		0·0010
			£0·5727
Add Profit and H.O. costs 10%			0·0573
			£0·6300

Gross labourer's rate: 63p per hour

With the addition of profit and H.O. costs to the variable overheads, a check can be made to ensure mathematical accuracy:

£0·6300 − £0·3750 = £0·2550 overheads
68% of £0·3750 = £0·2550 overheads

These figures differ for each contract but need calculating once only at the commencement of the costing system. The example above has shown the variable overheads for a labourer. Similar figures must be compiled for tradesmen and an average calculated as suggested in Chapter 4 (*Figure 4.3*).

Example 7.2 shows the breakdown of a tradesmen's rate based on a basic of £0·4396 per hour, graduated pension of £0·60 per week, tool money of £0·10 per week, wet time of 1% and importation based on 50% tradesmen on subsistence at £5·95 per week and 50% on travelling time of 6 h per week.

Example 7.2

Tradesmen's rate		£
Basic rate		0·4396
Plus rates		—
Tool money $\frac{0\cdot10}{50}$		0·0020
Attraction money		—
Stamps	£	
Holiday stamps, public	0·40	
annual	0·70	
National Insurance	0·90	
Graduated pension ave.	0·60	
S.E.T.	2·40	
	£5·00	

Assume 50 h week $= \dfrac{5\cdot00}{50}$ 　　　　　　　　　　　　　　　0·1000

Non-productive O/T $\dfrac{10 \times \frac{1}{2} \times 0\cdot4396}{50}$ 　　　　　　　　　　0·0440

Importation

Travelling $\dfrac{50\% \times 6\text{ h @ }0\cdot4396}{50}$ 　　　　　　　　　　　　0·0264

Lodging allowance $\dfrac{50\% \times 5\cdot95}{50}$ 　　　　　　　　　　　　0·0595

　　　　　　　　　　　　　　　　　　　　　　　　　£0·6715

Wet time 1%　　　　　　　　　　　　　　　　　　　　0·0067

Redundancy, basic　　　　　　0·4396
　　　　non-productive O/T　0·0440
　　　　　　　　　　　　　　─────
　　　　　　　　　　　　　　0·4836 × ¼%　　　　　0·0012
　　　　　　　　　　　　　　　　　　　　　　　　　─────
　　　　　　　　　　　　　　　　　　　　　　　　　£0·6794
Add Profit and H.O. costs 10%　　　　　　　　　　　　0·0679
　　　　　　　　　　　　　　　　　　　　　　　　　─────
　　　　　　　　　　　　　　　　　　　　　　　　　£0·7473
　　　　　　　　　　　　　　　　　　　　　　　　　═════

Gross tradesmen's rate: 75p/h

Expressed as a percentage, the tradesman's variable overheads are as shown in *Figure 7.1*.

BASIC	0.4396	100%
PLUS RATES	—	—
TOOL MONEY	0.0020	0.45%
ATTRACTION MONEY	—	—
HOLIDAY STAMPS	0.0220	5.00%
NATIONAL INSURANCE	0.0180	4.10%
GRADUATED PENSION	0.0120	2.74%
S.E.T.	0.0480	10.93%
NON-PRODUCTIVE O/T	0.0440	10.00%
IMPORTATION	0.0859	19.55%
G.M.U. & I.W.T. (WET TIME)	0.0067	1.52%
REDUNDANCY	0.0012	0.27%
PROFIT AND H.O. COSTS	0.0682	15.50%
	0.3080	70.06%

Figure 7.1

Based on an average of one labourer to one tradesman, the average percentages for all men on this contract are as shown in *Figure 7.2*.

BASIC RATE	LAB. £0.3750	TRADES £0.4396	AVERAGE £0.4073
	%	%	%
PLUS RATES	1.34	—	0.90
TOOL MONEY	—	0.45	
ATTRACTION MONEY	—	—	—
HOLIDAY STAMPS	5.87	5.00	5.44
NATIONAL INSURANCE	4.80	4.10	4.45
GRADUATED PENSION	2.66	2.74	2.70
S.E.T.	12.80	10.93	11.86
NON-PRODUCTIVE O/T	10.00	10.00	10.00
IMPORTATION	12.00	19.55	15.77
G.M.U. & I.W.T. (WET TIME)	2.98	1.52	2.25
REDUNDANCY	0.27	0.27	0.27
PROFIT AND H.O. COSTS	15.30	15.50	15.40
	68.02	70.06	69.04

Figure 7.2

Thus, if the total nett value or standard of measured work for a week is £1 200, the contract will also have earned 0·90% × £1 200 as the value or standard for plus rates and tool money for that week. Similarly, 5·44% × £1 200 is the standard for holiday stamps, 11·86% the standard for S.E.T., etc.

In the breakdowns shown travelling time, fares and lodging allowance have been grouped together under the heading 'Importation'. Holiday, National Insurance, graduated pension and S.E.T. stamps can also be grouped together under a heading 'Stamps', although the inclusion of a sliding scale of charges such as the graduated pension scheme tends to cloud an otherwise clear warning of 'too many men' given by a loss against standard, on, for example, S.E.T.

Exercises

1. Calculate gross rates for labourers and tradesmen based on current basics and express the variable overheads as average percentages of the average basic rate.

2. State the differences in approach to setting standards or budgets for fixed and for variable-type overheads.

8
FEED-BACK OF OUTPUT DATA

To be able to build up an adequate library of output standards, a costing system must be geared to recording the relevant historical data.

It was stated in Chapter 6 that the ideal period for recording output data is no less than from start to finish of the operation being studied. The costing system must therefore be arranged to collect measure of work done and totals of labour and plant expended for each operation being costed. This can be carried out at the completion of each contract, but the more preparatory work that has already been done the better. It is no easy task to filter through months, even years, of site costs in an attempt to abstract measurements and expenditure for like items that may be described quite differently by different personnel and yet could have been easily collected together under the same heading as work was being carried out.

To ensure that such collections of data are prepared as the work proceeds, the feed-back records can be incorporated into the costing system by using them as the collecting sheet for similar work by different gangs. This 'feed-back' sheet can also be an information sheet containing the relevant labour and plant standards, as illustrated in *Figure 8.1*.

It is possible to abstract task sheets or allocation sheets, etc., directly on to the feed-back record sheet, as shown in *Figure 8.2*. This procedure, however, is paper-consuming (continuation sheets may have to be written out each few weeks) and requires daily searching through an ever-increasing number of sheets to find each item of entry. It is therefore recommended only for small contracts and would be unsuitable for systems incorporating an incentive scheme, as the weekly sub-totals would have to be abstracted on to a separate form for bonus calculations similar to *Figure 9.1*. It is advisable before setting up a system of feed-back to decide on which items are to be recorded, what they are to include and what is to be

Feed-back of Output Data

CONTRACT									
FEED-BACK RECORD SHEET									
STANDARD VALUE	£0.15		£0.20		£0.30				
ITEM	Lay and joint 100 mm dia. S.G.W. pipes		ditto 150 mm		ditto 300 mm				
WEEK NO.	m	h	m	h	m	h			
15	30	10							
16	45	13	10	6					
	20	8	25	15					
17	35	13							
	28	10			10	7			
18			20	13	10	6			
19	14	6							

Figure 8.1

ignored, otherwise inconsistencies will occur from contract to contract. Feed-back of operations using plant should be further separated by types of plant. For example, trench excavation by JCB 3 and JCB 6 recorded together will in total give a meaningless output. At the end of a contract the feed-back records shown in *Figures 8.1* and *8.2* can be totalled and, with the addition of general comments applicable to the contract as a whole, the master records completed as illustrated in *Figure 6.1*.

If it is required to obtain outputs that reflect the possible rather than the actual then inefficiencies during the contract must be kept separate. This requires the system of labour and plant allocations (Chapter 5) to be sufficiently detailed to enable lost time to be isolated. Alternatively, outputs can be calculated weekly and those outside a permitted deviation from the current mean excluded from feed-back records as being questionable.

Where work study standards are used for production control then not only must output be recorded but also overall performances (B.S. 3138 : 1969 Clause A1024) of various trades and type of contract, in order that the data can be used for overall planning and estimating. It is optimistic to assume that 100 overall performance can be maintained even under incentive conditions.

CONTRACT _____

FEED-BACK RECORD SHEET

STANDARD VALUE	£0.20		£0.25		£0.40		£0.50		
ITEM	50 mm paving slabs in large areas		Ditto to decorative pattern		50 mm gravel paths hand rolled between paving slabs		250 mm x 50 mm p.c. edging incl. backing with concrete		
WEEK NO.	m^2	h	m^2	h	m^2	h	m^2	h	
28 M		16							
T		4		12					
W				16					
T		8						8	
F				8				8	
S				8					
SUB-TOTAL	66	28	70	44			10	16	
29 M				16					
T				16					
W				16					
T				16					
F						16			
S						8			
SUB-TOTAL			101	64	30	24			
30 M						8		8	
T						8		8	
W		16							
T		16							
F		16							
S						8			
SUB-TOTAL	89	48			26	24	11	16	

Figure 8.2

9
INCORPORATING A BONUS SYSTEM

The bookkeeping requirements of a bonus system are very similar to those of standard costing inasmuch as the expenditure for an operation or series of operations is compared with the target for that work, the difference, if a gain, being the amount of bonus earned. *Figures 9.1* and *9.2* show the calculation of a gang bonus, which, as with standard costs, can be computed in either hours or cash.

The terms of a bonus system may require a rate of payment for hours saved, either below or above the current labour rate. Targets are sometimes eased to make them more attractive but a reduction in the rate per hour saved invariably follows. On the other hand, a rate higher than the current labour rate may be used for payment of hours saved in an attempt to make savings on overheads by attracting the most efficient labour. In the example illustrated in *Figures 9.1* and *9.2* payment of the gang's savings is made at an agreed rate of 30p/h.

It can be quickly seen that by incorporating the bonus system into the standard cost duplication of bookkeeping is avoided. Bonus can be calculated as illustrated and the collected data of measurements and expenditure transferred to the feed-back collection sheets as suggested in Chapter 8. If collection of data for feed-back is not required, the information can be transferred directly to a standard cost statement, costing standards being substituted for bonus targets.

Where bonus is paid on an output achieved basis rather than an hours saved basis, with targets set proportionately more difficult, the same format can be used, the only difference being in the method of calculating payment. Achievement of the target usually entitles the operatives to a set amount of bonus, say one-third above basic wages. Failure to achieve the target decreases the amount of bonus proportionately, subject to any agreed minimum.

Where work study techniques are used for assessing targets,

Figure 9.1 in itself becomes a standard cost, comparing work study standards with actual costs for cost control purposes and producing a gang performance index for bonus payment purposes, as illustrated in *Figure 9.3*. Bonus can be related to performance either as a straight proportional scheme, which could show bonus payments of say 50% above basic pay at 100 performance, reducing to nil bonus at $66\frac{2}{3}$ performance, or alternatively by adjusting the lower

W/E 11/3/72												FORM NO. C/2
WEEK NO. 23												SHEET NO. 6
QTY.	UNIT	UNIT TARGET	TOTAL TARGET	DESCRIPTION OF WORK	M	T	W	T	F	S	S	TOTAL HOURS
25	m^2	2.2	55	Fix fwk. to cols.			27	9	8			44
10	m^2	1.8	18	Fix fwk. to beams						15		15
68	m^2	1.1	75	Fix fwk. to walls	27	18			16			61
45	m^2	0.7	31	Strip fwk. to walls				18	16			34
72	m^2	1.3	94	Make beam fwk.	18	18	18	18				72
45	m^2	0.4	18	Repair wall shutters						15		15
TOTAL TARGET: 291 HOURS									**TOTAL COST** 241 HOURS			

Figure 9.1

Incorporating a Bonus System

performance level; the amount paid at 100 performance or the average performance expected (allowed time) for a certain payment level, any type of payment scheme shown in *Figure 8* of B.S. 3138: 1969 can be produced.

Where the majority of tasks within a contract can be identified, pre-measured and targeted prior to that section of work commencing, a job card system may be set up showing on each

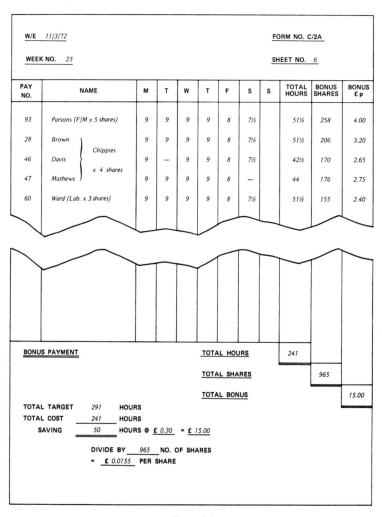

Figure 9.2

card the target hours or standard hours for that particular task. The total target for work done during one week is thus the addition of all completed job cards for that week set against actual times.

W/E 7/6/70 WEEK NO. 23													FORM NO. C/2 SHEET NO. 6		
QTY	UNIT	UNIT STANDARD	TOTAL STANDARD	DESCRIPTION OF WORK	M	T	W	T	F	S	S	TOTAL HOURS	GAIN ON STANDARD	LOSS ON STANDARD
25	m^2	1.8	45	Fix fwk. to cols			27	9	8			44	1	-
10	m^2	1.4	14	Fix fwk. to beams					15			15	-	1
68	m^2	0.8	54	Fix fwk. to walls	27	18		16				61	-	7
45	m^2	0.5	23	Strip fwk. to walls			18	16				34	-	11
72	m^2	1.0	72	Make beam fwk.	18	18	18	18				72	-	-
45	m^2	0.4	18	Repair wall shutters					15			15	3	-

TOTAL STANDARD HOURS 226   TOTAL ACTUAL HOURS 241

$$\frac{\text{STANDARD} \times 100}{\text{ACTUAL}} = 94 \text{ \% PERFORMANCE}$$

Figure 9.3

10
STANDARD COST EXAMPLE

The following example follows the basic principles of a standard cost from collection of site data (in this case by ganger's allocation sheet) through the bonus system and feed-back records to a weekly cost statement. The final summary sheet of trades and overheads shows also the 'to-date' picture of the contract, so trends in overheads, financial progress compared with programme and over-all gains and losses can be observed.

The example chosen illustrates a nett cost presented in cash. The advantages of costing in cash rather than hours are as follows.

1. Labour only sub-contractors are usually paid on a measure basis which is not directly related to the number of hours they have worked.
2. If an item of plant different from or additional to the one anticipated is used, a complicated calculation has to be carried out to convert hours value of one machine into hours value of another or additional machine. In cash the value remains constant. Where standards are taken from a standards library, however, the alteration of method or plant is more simple, since reference only has to be made to the library for the revised standard.
3. Many overheads can only be expressed in cash.
4. The impact made by the cost statement is psychologically more effective when shown in £ than in hours.

There are, however, two main disadvantages of cash records.

1. Allocation of domestic labour and plant to various operations commences in hours; thus it requires conversion into cash at some point in the system if the final statement is to be in cash.
2. Feed-back of output data requires to be in hours rather than

cash, otherwise continual and undesirable up-dating of cash standards is necessary. This means that if the final statement is in cash, then outputs must be converted back into hours for record purposes; however, the example avoids this additional work by collecting feed-back data before the hours are converted to cash, i.e. before the standard cost statement.

STANDARD COST EXAMPLE

Measurements for the week ending 26 April 1970, i.e. week 7, are as follows:

Excavate basement and load (22RB)	1 000 m³
Excavate bases and load	50 m³
Excavate manholes and drains and load	28 m³
Excavate oversite and load	10 m³
Blinding	7 m³
Concrete bases	13 m³
Concrete drains	4 m³
Concrete slab	6 m³
Unload bricks	14 000 No.
Unload cement	9 tonne

All excavations carted away by labour only sub-contractor.

C.A. quantity on lorry   1 300 m³

Bonus targets are as follows:

Excavate basement and load by 22RB and dragline (machine target)	15 m³/h
Excavate bases and load (hand)	3½ h/m³
Excavate manholes and drains and load (hand)	4 h/m³
Excavate oversite and load (hand)	4½ h/m³
Blinding	5 h/m³
Concrete bases	3½ h/m³
Concrete drains	5 h/m³
Concrete slab	6 h/m³
Unload bricks	1½ h/thousand
Unload cement	½ h/tonne
Erect hut (estimated)	18 h
Chainboy (estimated)	6 h

Standard Cost Example

Savings on bonus targets are to be paid out at 30p per hour saved for labour and £1·50 per hour saved for 22RB. The ganger and 22RB driver are to receive 1¼ shares per hour and labourers 1 share per hour, each gang to be bonused separately. The 22RB gang's bonus is based on machine hours saved and not on labour hours saved.

Plant on site during the week is as follows:

22RB (hourly rate)

10/7   Concrete mixer @ £8·00 week

5/3½   Mortar mixer @ £3·00 week

The 'to-date' summary at week 6 is shown in *Figure 10.1*.

Figures 10.2–10.7 show allocation sheets as written out by the ganger J. Green in charge of the general labouring gang. Absentees are included on the sheets but are shown as 'A'. *Figures 10.8–10.14* show similar sheets completed by C. Harris, the 22RB driver, allocating for himself, his banksman and his machine. The men and machines shown on these sheets must be checked daily against the time-books or clocking system and discrepancies queried.

As a second check that man-hours are correct, comparison should be made with the wage sheet totals illustrated in *Figure 10.15* which also shows the summary of wage sheet items required for cost

(*text continued on page 86*)

DESCRIPTION OF WORK	VALUE £	COST £	GAIN £	LOSS £
MEASURED WORK				
Excavator	91	145	2	56
Excavator (plant)	(84)	(116)	—	(32)
Concretor	10	11	—	1
Productive bonus	—	—	—	—
OVERHEADS				
Fixed Labour (prelims)	150	219	9	78
Plant	(16)	(16)	—	—
Variables (on-costs)	134	183	13	62
SUB-LET				
Excavator	44	36	8	—
	529	726	32	229

Figure 10.1

ANY FIRM & SON LTD.											FORM NO. C/1
DAILY TIME ALLOCATION SHEET											

WAGES NUMBER	13	15	16	17	18	19	21	23	Staff			
NAME	Parish	Brown	Ovenden	Clews	Challis	Jenner	Eaton	Davis	Green			TOTALS
Dig bases	9		9	9					2			29
Dig drains		4										4
Blinding to drains		2							2			4
Cleaning out base that had fallen in after rainstorm		3			3		3	3	2			14
Unload cement					1	1	1	1	1			5
Unload bricks					2	2	2	2				8
Oversite excavation for slab					3			1	1			5
Dig manhole						5	2	2				9
Discuss office erection with Agent									1			1
	9	9	9	9	9	8	8	9	9			79

CONTRACT:
GANGER: *J.Green*
DATE: *Monday 20th April, 1970*
WORKS MANAGER OR GENERAL FOREMAN: *K.T.*

Figure 10.2

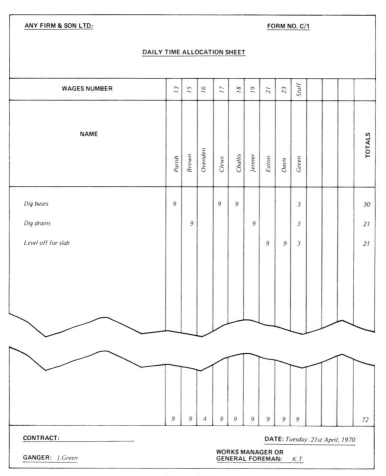

Figure 10.3

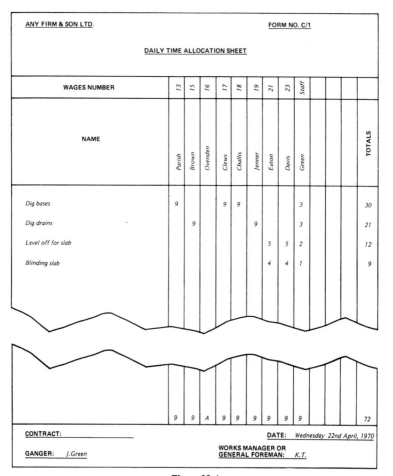

Figure 10.4

ANY FIRM & SON LTD. FORM NO. C/1

DAILY TIME ALLOCATION SHEET

WAGES NUMBER	13	15	16	17	18	19	21	23	Staff			
NAME	Parish	Brown	Ovenden	Clews	Challis	Jenner	Eaton	Davis	Green			TOTALS
Dig bases	8			8	8				2			26
Blind bases	1	3		1	1	3	3		1			13
Unload bricks								2	2			4
Helping to put up office						6	6					12
Assisting Engineer with setting out									4			4
Dig drains		5						8	1			14
	9	8	A	9	9	9	9	10	10			73

CONTRACT: 　　　　　　　　　　DATE: Thursday 23rd April, 1970
GANGER: J. Green 　　　　WORKS MANAGER OR GENERAL FOREMAN: K.T.

Figure 10.5

ANY FIRM & SON LTD.												FORM NO. C/1

DAILY TIME ALLOCATION SHEET

WAGES NUMBER	13	15	16	17	18	19	21	23	Staff			
NAME	Parish	Brown	Ovenden	Clews	Challis	Jenner	Euton	Davis	Green			TOTALS
Concrete to drains						4	4	4	3			15
Dig bases	8			8	4							20
Concrete bases 2B & 2C		4				4		4	3			15
Helping to put up office					4							4
Dig drains		4					4		2			10
	8	8	A	8	8	8	8	8	8			64

CONTRACT: **DATE:** Friday 24th April, 1970

GANGER: J.Green   **WORKS MANAGER OR GENERAL FOREMAN:** K.T.

Figure 10.6

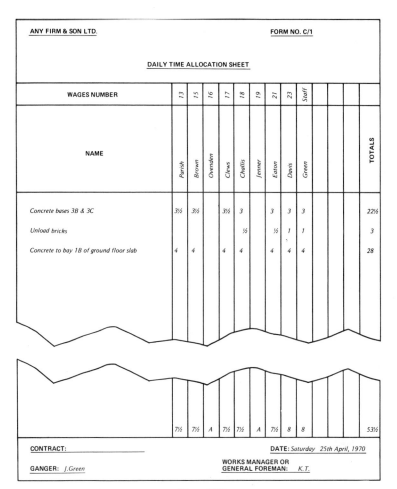

Figure 10.7

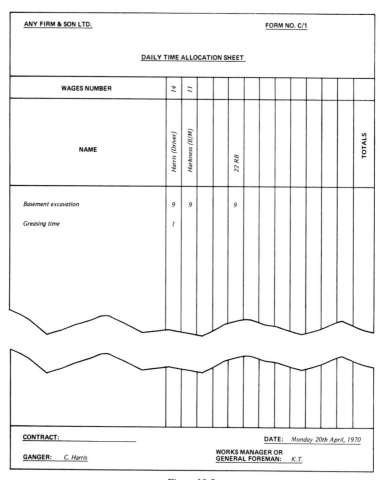

Figure 10.8

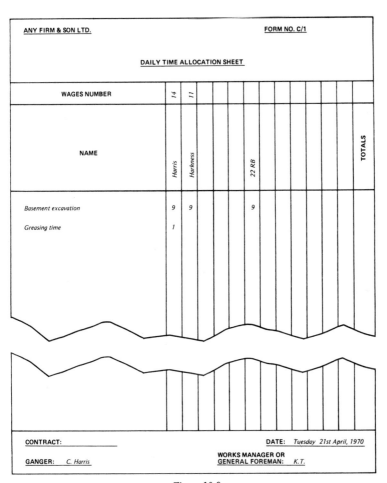

Figure 10.9

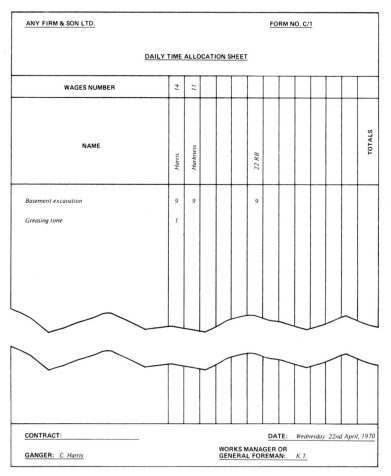

Figure 10.10

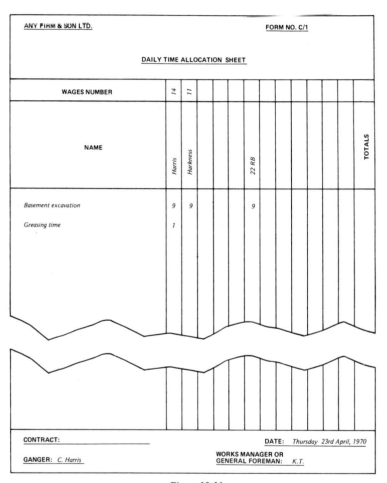

Figure 10.11

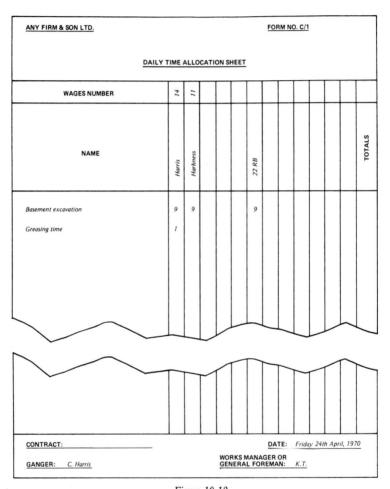

Figure 10.12

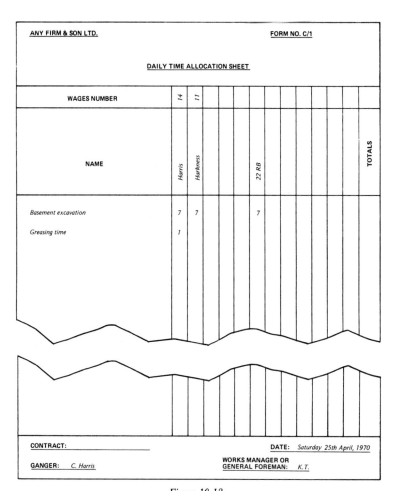

Figure 10.13

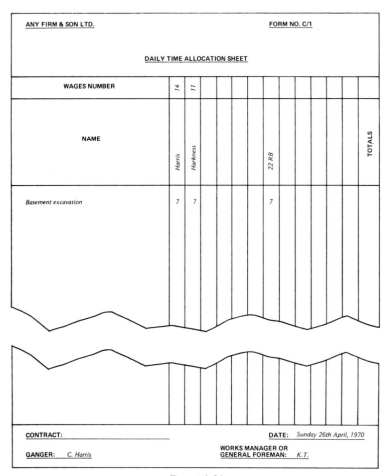

Figure 10.14

ANY FIRM & SON LTD.		FORM NO.W/2	
TIME & WAGES WEEKLY COST SUMMARY			
BREAKDOWN OF HOURS		DIRECT LABOUR HOURS	WORKING STAFF HOURS
NORMAL WORKING HOURS		368	40
OVERTIME		110½	13
NON-PRODUCTIVE OVERTIME		53½	5¾
TRAVELLING TIME		24	—
MAINTENANCE TIME		6	—
INCLEMENT WEATHER TIME		—	—
GUARANTEED MAKE-UP		—	—
		562	58¾
ADD WORKING STAFF		58¾	
TOTAL HOURS		620¾	

BREAKDOWN OF WAGES	DIRECT LABOUR £ p	WORKING STAFF £ p
FLAT RATE WAGES	223.65	26.45
EXTRAS OVER RATE	—	—
BONUS (PREVIOUS WEEK)	—	—
TOOL MONEY	—	—
EXPENSES	3.00	—
LODGING ALLOWANCE	12.00	6.00
STATE PENSION	6.50	0.75
SELECTIVE EMPLOYMENT TAX	24.00	2.40
HOLIDAY STAMPS	11.00	—
NATIONAL INSURANCE CONTRIBUTION	8.96	0.90
REDUNDANCY PAYMENTS	—	—
	289.11	36.50
ADD WORKING STAFF	36.50	
TOTAL WAGES	£325.61	

CONTRACT _____ W/E 26th April, 1970

Figure 10.15

W/E 26th April, 1970										FORM NO.		C/2
WEEK NO. 7										SHEET NO.		1

QTY.	UNIT	UNIT TARGET	TOTAL TARGET HOURS	DESCRIPTION OF WORK	M	T	W	T	F	S	S	TOTAL ACTUAL HOURS
50	m^3	3½	175	Excav. bases	43	30	30	26	20			149
28	m^3	4	112	Excav. drains	13	21	21	14	10			79
10	m^3	4½	45	Excav. oversite	5	21	12					38
7	m^3	5	35	Blinding	4		9	13				26
4	m^3	5	20	Concrete drains					15			15
13	m^3	3½	45½	Concrete bases					15	22½		37½
6	m^3	6	36	Concrete floor slab						28		28
14 000	No.	1½ per thou	21	Unload bricks	8		4			3		15
9	tonne	½	4½	Unload cement	5							5
			18	Erect office	1			12	4			17
			6	Chainman				4				4

TOTAL TARGET: 518 HRS.   TOTAL COST: 413½ HRS.

Figure 10.16

W/E 26th April, 1970								FORM NO. C/2A		
WEEK NO. 7								SHEET NO. 1		

PAY NO.	NAME	M	T	W	T	F	S	S	TOTAL HOURS	BONUS SHARES	BONUS £ p
13	Parish	9	9	9	9	8	7½		51½	52	3.80
15	Brown	9	9	9	8	8	7½		50½	51	3.73
16	Ovenden	9	—	—	—	—	—		9	9	0.65
17	Clews	9	9	9	9	8	7½		51½	52	3.80
18	Challis	9	9	9	9	8	7½		51½	52	3.80
19	Jenner	8	9	9	9	8	—		43	43	3.15
21	Eaton	8	9	9	9	8	7½		50½	51	3.73
23	Davis	9	9	9	10	8	8		53	53	3.87
Staff	Green (Ganger)	9	9	9	10	8	8		53	66	4.82

BONUS PAYMENT

TOTAL HOURS 413½
TOTAL SHARES 429
TOTAL BONUS 31.35

TOTAL TARGET 518 HOURS
TOTAL COST 413½ HOURS
SAVING 104½ HOURS @ £ 0.30 = £ 31.35

DIVIDE BY 429 NO. OF SHARES
= £ 0.073 PER SHARE

Figure 10.17

W/E 26th April, 1970												FORM NO. C/2
WEEK NO. 7												SHEET NO. 2
QTY.	UNIT	UNIT TARGET	TOTAL TARGET	DESCRIPTION OF WORK	M	T	W	T	F	S	S	TOTAL HOURS
				Excav. basement	18	18	18	18	18	14	14	118
				Greasing time (MT)	1	1	1	1	1	1		6
				(22RB)								
1000	m³	1/15	67	Excav. basement	9	9	9	9	9	7	7	59

TOTAL TARGET 67 HRS.   TOTAL COST 59 HRS.

Figure 10.18

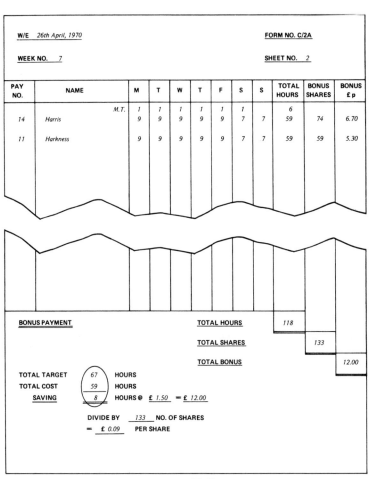

Figure 10.19

Standard Cost Example

CONTRACT							FORM NO. C/3		

FEED-BACK RECORD SHEET

Excavator

REF.	13/A				13/B			25/P – 26/F		
VALUE	L. £0.40		(P. £0.50)		L. £0.50	(P. £0.20)		L. £0.50	(P. £0.50)	
DESC.	Excavate o/s.				Excavate bases			Excavate drains and manholes		
WEEK	B.O.Q. 830	£1.25	£1.50	£0.40	B.O.Q. 650	£1.50	£0.40	B.O.Q. 250	£1.50	£0.40
	m³	Hours Drott	Hours 22 RB	Hours Lab.	m³	Hours 22 RB	Hours Lab.	m³	Hours 22 RB	Hours Lab.
5	20	10		6 20						
6	1∅0	40	4	10 84	70	2 10	200			
	120	50	4	120	70	12	200			
7	10	–	–	38	50	–	149	28	–	79
	130	50	4	158	120	12	349	28	–	79

Figure 10.20

control of variable overheads. In practice this summary can be a stumbling block to the whole system, as wages may well be calculated by remote computer or by clerks who have other duties, thus not providing the summary until late in the week. However, the cost statement for productive work can be prepared and corrective action taken where indicated prior to the addition of the wage sheet items. Overheads are by their nature a more long-term problem and do not therefore have the same urgency as the calculation of bonuses and preparation of the cost statement for productive work.

Figures 10.16 and *10.18* are abstracts of work from J. Green's and C. Harris's allocation sheets, together with the measure of work and bonus targets. The total target hours and cost hours are carried forward to *Figures 10.17* and *10.19* for calculation and sharing of any bonus due. In order that a deadline for calculating bonus payments can be met it is essential to abstract allocation sheets daily and pre-measure as much work as possible, so that on measurement day (usually Monday) only the weekend's allocation need be entered

Standard Cost Example

CONTRACT					FORM NO. C/3		
			FEED-BACK RECORD SHEET				
Excavator					Sub-Let		
REF.	13/C				13/D		
VALUE	L.£0.08	P.£0.10			Gross L & P. £0.15 m³		
DESC.	Excavate basement				Cart away		
WEEK	B.O.Q. 3 015	£1.50	£0.40		B.O.Q. 4495	£0.10	
	m³	Hours 22 RB	Hours Lab.		m³	Sub-Let m³	
5					20	24	
6	100	20	2 40		270	340	
	100	20	42		290	364	
7	1 000	59	6 118		1 088	1 300	
	1 100	79	166		1 378	1 664	

Figure 10.21

up, and the odd few measurements taken. By adhering to these principles it is possible to calculate bonuses with only a few hours of the start of a new week, thus avoiding the unsatisfactory practice of paying bonuses one week after wages, i.e. two weeks after the bonus has been earned. Greater incentive effect is obtained by publishing bonus earnings or gang performance levels as soon as they have been calculated.

The next stage is to collect similar work together as shown in *Figures 10.20–10.24* and, as they become available, the amounts for *Figures 10.25–10.27*. The unit values or standards have been noted on these sheets for cashing out the cost statement as illustrated in *Figures 10.28* and *10.29* and the final summary as illustrated in *Figure 10.30*.

The order of working through the various clerical procedures within the system has been so arranged that information is available in the most suitable units with the minimum of effort: first, the calculation of bonus in hours; second, the recording of feed-back information again in hours and tied into the system so that it cannot be left behind; third, the production of the cost statement and summary in cash. By this method the advantages of working both in hours and in cash are utilised.

CONTRACT									FORM NO. C/3		
Concretor					FEED-BACK RECORD SHEET						
REF.	14/A–D & 26/S		14/E		14/H		26/T		14/J		
VALUE	L. £2.00		L. £1.50		L. £2.25		L. £1.80		L. £2.60		
DESC.	Blinding		Bases		6" Floor slab		Surround to drains		Columns		
WEEK	B.O.Q. 121	£0.40	B.O.Q. 229	£0.40	B.O.Q. 822	£0.40	B.O.Q. 193	£0.40	B.O.Q. 198	£0.40
	m^3	Hours Lab.	m^3	Hours Lab.	m^3	Hours Lab.	m^3	Hours Lab.	m^3	Hours Lab.
6	5	28								
	<u>5</u>	<u>28</u>								
7	7	26	13	37	6	28	4	15		
	<u>12</u>	<u>54</u>	<u>13</u>	<u>37</u>	<u>6</u>	<u>28</u>	<u>4</u>	<u>15</u>		

Figure 10.22

CONTRACT						FORM NO. C/3	

FEED-BACK RECORD SHEET

Labour Prelims.

REF.	Bill 1		Bill 1		Bill 1		Bill 1	
VALUE	£10.00 per week for 20 weeks		£5–£20 per week until £500		£100 erect £25 dismantle		£40 erect. £20 dismantle	
DESC.	Chainman		Unload materials		Temporary buildings		Mixer set-up	
WEEK	Value used	£0.40	Value used	£0.40	Value used	£0.40	Value used	£0.40
	£	Hours Lab.	£	Hours Lab.	£	Hours Lab.	£	Hours Lab.
5	10	12	5	20	40	200	40	90
6	10	27	5	17	40	180		
	_20_	_39_	_10_	_37_	_80_	_380_	_40_	_90_
7	10	4	5	20	20	17		
	_30_	_43_	_15_	_57_	_100_	_397_	_40_	_90_

Figure 10.23

CONTRACT						FORM NO. C/3		
Plant Prelims.			FEED-BACK RECORD SHEET					
REF.		*Bill 1*			*Bill 1*			
VALUE	£8.00 per week for 30 weeks				£4.00 per week for 40 weeks			
DESC.		*Concrete Mixer*			*Mortar Mixer*			
WEEK	Value used	£8.00			Value used	£3.00		
	£	10/7 weeks			£	5/3½ weeks		
5	8	1						
6	8	1						
	<u>16</u>	<u>2</u>						
7	8	1			4	1		
	<u>24</u>	<u>3</u>			<u>4</u>	<u>1</u>		

Figure 10.24

CONTRACT				FORM NO. C/3				
			FEED-BACK RECORD SHEET					
On-Costs								
REF.	Bill rates		Bill rates		Bill rates			
VALUE	10.00%		2.25%		15.77%			
DESC.	Non-Prod. Overtime		Wet Time		Importation of Labour			
WEEK		£0.40		£0.40	£0.40	£	£	
		Hours Lab.		Hours Lab.	Travel Time	Exps.	Sub-sistence	
5		25		14	10	1	6	
6		23		20	15	1	18	
		<u>48</u>		<u>34</u>	<u>25</u>	<u>2</u>	<u>24</u>	
7		59		—	24	3	18	
		<u>107</u>		<u>34</u>	<u>49</u>	<u>5</u>	<u>42</u>	

Figure 10.25

CONTRACT							FORM NO. C/3	
On-Costs			FEED-BACK RECORD SHEET					
REF.	*Bill rates*		*Bill rates*		*Bill rates*		*Bill rates*	
VALUE	2.70%		11.86%		5.44%		4.45%	
DESC.	Graduated Pension		S.E.T.		Holiday Stamps		Employer's Nat. Ins.	
WEEK		£		£		£		£	
5		8		34		16		14	
6		12		51		24		21	
		<u>20</u>		<u>85</u>		<u>40</u>		<u>35</u>	
7		7		26		11		10	
		<u>27</u>		<u>111</u>		<u>51</u>		<u>45</u>	

Figure 10.26

CONTRACT						FORM NO. C/3	
On–Costs			**FEED-BACK RECORD SHEET**				
REF.	Bill rates		Bill rates	Bill rates		Bill rates	
VALUE	0.90%		Nil	0.54% during second half of contract		N/A Rise & Fall Clause applies	
DESC.	Plus Rates		Attraction Money	Redundancy		National Increases	
WEEK		£	£		£	£	
5		—	—		—	—	
6		—	—		—	—	
7		—	—		—	—	

Figure 10.27

		FORM NO. C/4	
CONTRACT		W/E 26th April, 1970	
SHEET NO. 1		WK NO. 7 of 75	
	COST CONTROL STATEMENT		

DESCRIPTION OF WORK	VALUE £	COST £	GAIN £	LOSS £
Excavator				
Oversite excavation	4	15	—	11
Oversite excavation	(5)	(—)	(5)	(—)
Excavate bases	25	60	—	35
Excavate bases	(10)	(—)	(10)	(—)
Excav. drains & manholes	14	32	—	18
Excav. drains & manholes	(14)	(—)	(14)	(—)
Excavate basement	80	50	30	—
Excavate basement	(100)	(89)	(11)	(—)
TOTAL EXCAVATOR	123	157	30	64
	(129)	(89)	(40)	(—)
Concretor				
Blinding	14	10	4	—
Concrete bases	20	15	5	—
Concrete floor slab	13	11	2	—
Concrete surround to drains	7	6	1	—
TOTAL CONCRETOR	54	42	12	—
SUB-LET				
Excavator				
Cart away	163	130	33	—

Figure 10.28

		FORM NO. C/4
CONTRACT:		W/E 26th April, 1970
SHEET NO. 2		WK NO. 7 of 75
	COST CONTROL STATEMENT	

DESCRIPTION OF WORK	VALUE £	COST £	GAIN £	LOSS £
Labour Prelims.				
Chainman	10	2	8	—
Unload Materials	5	8	—	3
Temporary Buildings	20	7	13	—
TOTAL LABOUR PRELIMS.	35	17	21	3
Plant Prelims.				
Concrete Mixer	(8)	(8)	—	—
Mortar Mixer	(4)	(3)	(1)	—
TOTAL PLANT PRELIMS.	(12)	(11)	(1)	—
Site On–Costs				
Non-Prod. Overtime	21	24	—	3
Wet Time	5	—	5	—
Importation of Labour	33	30	3	—
Graduated Pension	6	7	—	1
S.E.T.	25	26	—	1
Holiday Stamps	11	11	—	—
Employer's Nat. Ins.	9	10	—	1
Plus Rates	2	—	2	—
Attraction Money	—	—	—	—
Redundancy	—	—	—	—
TOTAL ON-COSTS	112	108	10	6

Figure 10.29

CONTRACT:					FORM NO. C/5			
					W/E 26th April, 1970			
					WK NO. 7 OF 75			

COST CONTROL SUMMARY SHEET

	THIS WEEK				TO DATE			
	VALUE	COST	GAIN	LOSS	VALUE	COST	GAIN	LOSS
MEASURED WORK								
Excavator	123	157	30	64	214	302	32	120
Excavator (Plant)	(129)	(89)	(40)	—	(213)	(205)	(40)	(32)
Concretor	54	42	12	—	64	53	12	1
Productive bonus	—	43	—	43	—	43	—	43
TOTAL MEASURED LABOUR	177	242	42	107	278	398	44	164
TOTAL MEASURED PLANT	(129)	(89)	(40)	—	(213)	(205)	(40)	(32)
TOTAL MEASURED	306	-331	82	107	491	603	84	196
OVERHEADS								
FIXED LABOUR PRELIMS.	35	17	21	3	185	236	30	81
PLANT PRELIMS.	(12)	(11)	(1)	—	(28)	(27)	(1)	—
On-costs on £2/2 value	112	108	10	6	246	291	23	68
TOTAL OVERHEADS	159	136	32	9	459	554	54	149
SUB-LET								
Excavator	163	130	33	—	207	166	41	—
TOTAL SUB-LET	163	130	33	—	207	166	41	—
GRAND TOTAL	628	597	147	116	1157	1323	179	345

GAIN/LOSS THIS WEEK £ 31 = 5.0 %

GAIN/LOSS TO DATE £ 166 = 14.3 %

Figure 10.30

Standard Cost Example

A check on mathematical error on the expenditure side of the standard cost is provided by a simple reconciliation of wages sheets with total labour costs shown in the standard cost statement.

Cost Reconciliation

From standard cost statement

	£
Measured work	242
Labour prelims.	17
On-costs	108
	£367

From wages sheets

	£
Grand cash total	326
Deduct previous week's bonus	—
	£326
Add this week's bonus	43
	£369

Similar checks can be carried out on plant and sub-let work if a weekly summary of these costs is kept apart from the standard cost. It is, however, the labour element of the calculations that is more prone to error because of the numerous calculations involved. Care must nevertheless be taken to include all plant and all labour only sub-contractors in the cost if plant is not finally reconciled in the same manner.

A further check on mathematical accuracy within the cost statement itself can be made by the simple addition of values and losses compared with costs and gains. For example, grand total for week 7:

	£		£
Value	628	Cost	597
Loss	116	Gain	147
	£744		£744

This particular check can be frequently carried out during the compilation of the standard cost as the rule applies equally to individual operations, trade totals, etc. For example, total plant listed under excavator:

	£		£
Value	129	Cost	89
Loss	—	Gain	40
	£129		£129

Exercise

1. Illustrate your proposals for a standard cost system for a large building firm dealing in hospitals, factories, schools, etc., where the incentive scheme is based on a straight plus rate addition to hourly rates and does not therefore require provision for calculation within the costing system. The statement is to be presented in hours and feed-back of data is to be recorded by a central office on receipt of the standard cost statement.

2. Design a bonus calculation sheet where work study standards are used as bonus targets. Suggest a means of extending the bonus system to produce a standard cost.

11

STANDARD COST EXERCISE

Exercises

1. The following task sheets show the work programmed for each day of week 8, and the amount of time spent on each operation. Using the blank forms given and the feed-back sheets previously used for week 7, calculate bonus earnings and prepare a cost statement for week 8.

Measurements for week ending 3rd May, 1970 are as follows:

Excavate basement and load (22RB)	1 140 m³
Excavate bases and load	30 m³
Excavate drains and load	25 m³
Excavate oversite and load	20 m³
Blinding	9 m³
Concrete bases	12 m³
Concrete slab	11 m³
Unload bricks	3 000 No.

All excavations carted away by labour only sub-contractor.

        C.A. quantity on lorry   1 500 m³

Bonus targets are as for previous weeks with the chainboy estimated again at 6 hours and erecting partitions at 18 hours.

2. Design a control system based on work studied standard times to produce a standard cost statement and an incentive scheme that will pay bonus at the rate of $33\frac{1}{3}\%$ above basic at 100 operator performance (B.S. 3138 : 1969 Clause A1022) reducing in steps of 10 performance down to nil bonus at 50 performance. Work through the system using week 8 measurements and the following standard times:

Excavate basement and load by 22RB and dragline
 (standard time includes for banksman) 0·08 h/m³
Excavate bases and load (hand) 2·0 h/m³
Excavate drains and load (hand) 2·5 h/m³
Excavate oversite and load (hand) 3·0 h/m³
Blinding 3·5 h/m³
Concrete bases 2·0 h/m³
Concrete slab 4·0 h/m³
Unload bricks 1·0 h/thousand
Chainboy estimated 4 h
Erect partitions estimated 12 h

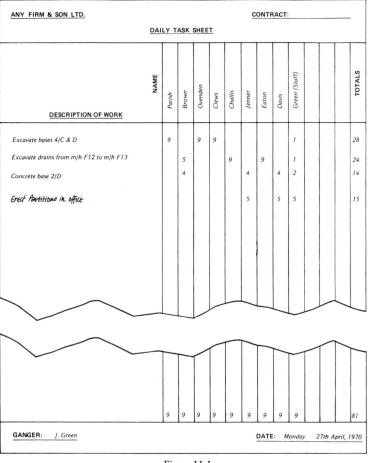

Figure 11.1

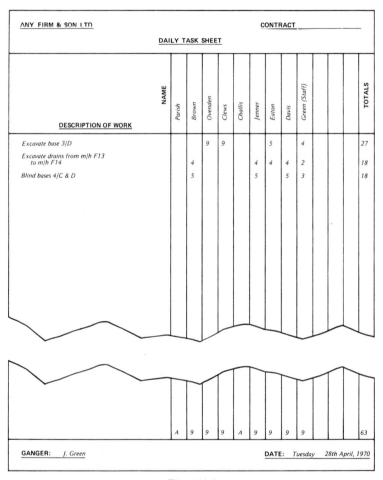

Figure 11.2

ANY FIRM & SON LTD.								CONTRACT			
DAILY TASK SHEET											
DESCRIPTION OF WORK	NAME	Parish	Brown	Overiden	Clews	Challis	Jenner	Eaton	Davis	Green (Staff)	TOTALS
Oversite excavation for floor slab		4	9	9			4	4	4	4	38
Concrete bases 4/C & D		5					4	4	4	4	21
Blind floor slab bay 1/C											
Unload bricks							1	1	1	1	4
		A	9	9	9	A	9	9	9	9	63

GANGER: J. Green **DATE:** Wednesday 29th April, 1970

Figure 11.3

ANY FIRM & SON LTD.									CONTRACT		
		DAILY TASK SHEET									
DESCRIPTION OF WORK	NAME										TOTALS
	Parish	Brown	Ovenden	Clews	Challis	Jenner	Eaton	Davis	Green (Staff)		
Excavate drains from m/h F14 to m/h F15			9	9	9		7	3			37
Blind floor slab bay 1/C						3	3	2	1		9
Excavate floor slab bay 2/C	9	9				6	6				30
Chainman to Engineer									6		6
	9	9	9	9	9	9	9	9	10		82
GANGER: J. Green							DATE: Thursday 30th April, 1970				

Figure 11.4

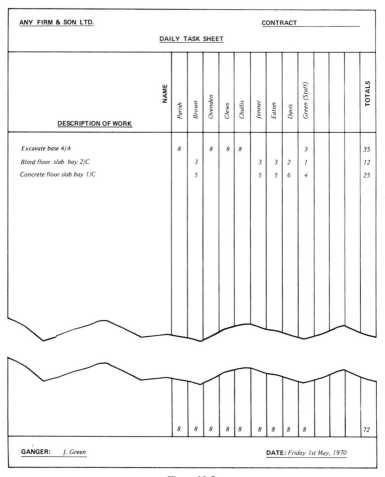

Figure 11.5

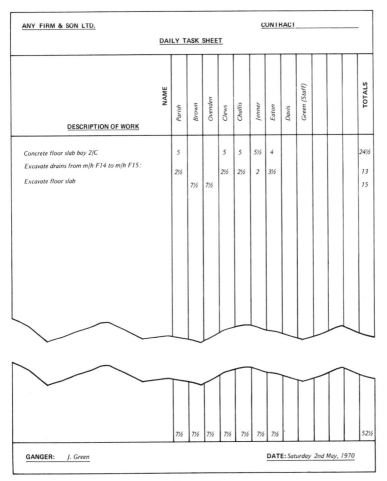

Figure 11.6

ANY FIRM & SON LTD.					CONTRACT				
DAILY TASK SHEET									
DESCRIPTION OF WORK	NAME Harris	Harkness (B/M)		22 RB					TOTALS
Monday									
Basement excavation	9	9		9					18
Greasing time	1								+1
Tuesday									
Basement excavation	9	9		9					18
Greasing time	1								+1
Wednesday									
Basement excavation	9	9		9					18
Greasing time	1								+1
Thursday									
Basement excavation	9	8		9					17
Greasing time	1								+1
Friday									
Basement excavation	8	8		8					16
Greasing time	1								+1
Saturday									
Basement excavation	7½	7½		7½					15
Greasing time	1								+1
Sunday									
Basement excavation	7½	7½		7½					15
	59 +6	58							117 +6
GANGER: C. Harris						DATE: W/E 3rd May, 1970			

Figure 11.7

ANY FIRM & SON LTD.		FORM NO. W/2
TIME & WAGES WEEKLY COST SUMMARY		

BREAKDOWN OF HOURS	DIRECT LABOUR HOURS	WORKING STAFF HOURS
NORMAL WORKING HOURS	368	40
OVERTIME	110	12½
NON-PRODUCTIVE OVERTIME	53¾	5¼
TRAVELLING TIME	24	—
MAINTENANCE TIME	6	—
INCLEMENT WEATHER TIME	—	—
GUARANTEED MAKE UP	—	—
	561¾	57¾
ADD WORKING STAFF	57¾	
TOTAL HOURS	619½	

BREAKDOWN OF WAGES	DIRECT LABOUR £ p	WORKING STAFF £ p
FLAT RATE WAGES	223.50	26.00
EXTRAS OVER RATE	—	—
BONUS (PREVIOUS WEEK)	—	—
TOOL MONEY	—	—
EXPENSES	3.00	—
LODGING ALLOWANCE	12.00	6.00
STATE PENSION	6.50	0.75
SELECTIVE EMPLOYMENT TAX	24.00	2.40
HOLIDAY STAMPS	11.00	—
NATIONAL INSURANCE CONTRIBUTION	8.96	0.90
REDUNDANCY PAYMENTS	—	—
	288.96	36.05
ADD WORKING STAFF	36.05	
TOTAL WAGES	325.01	

CONTRACT   W/E   3rd May, 1970

Figure 11.8

| W/E _____ | | | | | | | | | | | | FORM NO. C/2 |
| WEEK NO. _____ | | | | | | | | | | | | SHEET NO. _____ |

QTY.	UNIT	UNIT TARGET	TOTAL TARGET	DESCRIPTION OF WORK	M	T	W	T	F	S	S	TOTAL HOURS

TOTAL TARGET _____ HRS.   **TOTAL COST** _____ HRS.

Figure 11.9

PAY NO.	NAME	M	T	W	T	F	S	S	TOTAL HOURS	BONUS SHARES	BONUS £ p

W/E _____ FORM NO. C/2A

WEEK NO. _____ SHEET NO. _____

BONUS PAYMENT

TOTAL HOURS

TOTAL SHARES

TOTAL BONUS

TOTAL TARGET _____ HOURS
TOTAL COST _____ HOURS
　SAVING _____ HOURS @ £ _____ = £ _____
　　　　DIVIDE BY _____ NO. OF SHARES
　　　= £ _____ PER SHARE

Figure 11:10

QTY.	UNIT	UNIT TARGET	TOTAL TARGET	DESCRIPTION OF WORK	M	T	W	T	F	S	S	TOTAL HOURS

W/E _____ WEEK NO. _____ FORM NO. C/2 SHEET NO. _____

TOTAL TARGET _____ HRS.   TOTAL COST _____ HRS.

Figure 11.11

PAY NO.	NAME	M	T	W	T	F	S	S	TOTAL HOURS	BONUS SHARES	BONUS £ p

W/E _____
WEEK NO. _____
FORM NO. C/2A
SHEET NO. _____

BONUS PAYMENT

TOTAL HOURS
TOTAL SHARES
TOTAL BONUS

TOTAL TARGET _____ HOURS
TOTAL COST _____ HOURS
SAVING _____ HOURS @ £ ____ = £ ____

DIVIDE BY _____ NO. OF SHARES
= £ _____ PER SHARE

Figure 11.12

| CONTRACT | | | W/E | |
| SHEET NO. | | | WK NO. | |

COST CONTROL STATEMENT

DESCRIPTION OF WORK	VALUE £	COST £	GAIN £	LOSS £

Figure 11.13

CONTRACT			W/E		
SHEET NO.			WK NO.		
COST CONTROL STATEMENT					
DESCRIPTION OF WORK	VALUE £	COST £	GAIN £	LOSS £

Figure 11.14

CONTRACT					FORM NO. C/5 W/E WK NO. OF			
COST CONTROL SUMMARY SHEET								
	THIS WEEK				TO DATE			
	VALUE	COST	GAIN	LOSS	VALUE	COST	GAIN	LOSS
MEASURED WORK								
TOTAL MEASURED LABOUR								
TOTAL MEASURED PLANT								
TOTAL MEASURED								
OVERHEADS								
FIXED LABOUR PRELIMS.								
PLANT PRELIMS.								
ON-COSTS								
TOTAL OVERHEADS								
SUB-LET								
TOTAL SUB-LET								
GRAND TOTAL								

GAIN/LOSS THIS WEEK £   =   %
GAIN/LOSS TO DATE   £   =   %

Figure 11.15

12

COSTING SUNDRY ITEMS

SMALL ITEMS OF LABOUR

In order that a mathematical balance can be achieved and also to ensure that all corners of expenditure are being controlled, it is necessary to include all operations and overheads, no matter how small, in the standard cost system. It is not necessary, however, to keep every operation under a separate heading.

The cost of small items and their respective contributions to the contract's earnings or standard value can be:

1. Grouped together under trade headings, the value being either (*a*) physically measured against known standards, (*b*) related to the appropriate measured work either as a percentage or a value per unit of measurement, or (*c*) a previously calculated lump sum allowance each week, or
2. Charged against the appropriate measured work, the standard having previously been increased to allow for these small items. (*Note.* Care must be taken not to contaminate feed-back information by charging differing sundry labours, etc., against measured work, thus casting doubt as to what is included.)

Small items that are difficult to allocate, e.g. sundry labours, etc., are best charged against the appropriate measured work, their value having little effect on the standards. Sundry items, however, such as shuttering to nibs and pockets, are best grouped together as suggested in (1) above.

SMALL ITEMS OF PLANT

Light plant can be costed as a grouped item under the heading of 'Plant overheads', expenditure being compared with a budgeted weekly standard or value.

It is not, however, necessary to add together the cost of numerous similar items of plant each week in order to keep a record of expenditure for light plant. Revised totals can be calculated by adding newly hired items or deducting off-hire items as shown in *Figure 12.1*.

CREDITS

Where a profit and loss type of standard costing system is being operated, a contract may well be paid for work not actually carried out, as, for example, with planking and strutting, formwork to bases, etc. If costs have been incurred against this item, e.g. additional working space, etc., then the item can usually be costed in

DATE	LATRINES @ 5p/wk £	THEODOLITE @ 10p/wk £	LEVEL @ 8p/wk £	HAMMER @ 12p/wk £	SAW BENCH @ 45p/wk £	TOTAL £
5/3/70	0.05	0.10	0.08		0.45	0.68
12/3/70				0.12		0.80
19/3/70			0.08			0.72
26/3/70						0.72
3/5/70	0.05					0.77
10/5/70			0.08	0.12		0.73

Figure 12.1

the normal way. If, however, no expenditure has occurred or it is impossible to identify that expenditure, then the value of the operation can be shown as a credit to the contract and subsequently a straight gain. This should, of course, be indicated as a credit on the cost statement, as illustrated in *Figure 4.4* under 'Formwork to bases'.

STOP-ENDS

By far the simplest way of dealing with stop-ends is to extract the standard for make, fix and strip stop-ends from concreting rates or concreting standards and from then on treat stop-ends as a measurable and allocatable item, as would usually be done in a measured incentive scheme. To mix stop-end and concrete standards in the costing system reduces the accuracy of control, as ups and downs may occur in concreting expenditure because of more or less time spent during a certain week in fixing stop-ends.

DAYWORKS

Time spent on daywork operations can be recorded in the same way as normal measured work is recorded, i.e. by task sheets or ganger's allocation sheets, etc. Comparison can then be made by either of the following methods.

1. By adding or deducting a constant margin to the expenditure. For example, if daywork on labour is priced at $+100\%$, of which 10% is for planned profit and Head Office charges, and it is known that variable overheads are costing 75% above nett, then dayworks must be making 15% gain. Therefore, if the cost of daywork during one week was £100, the appropriate value would be £100 + 15% = £115, i.e. £15 gain. This is a rule-of-thumb method and relies on the variable overheads percentage remaining reasonably constant. It is, however, sufficiently accurate to complete the cost statement provided that the daywork does not form an appreciable percentage of the contract.
2. Where the extent of daywork is significant, it is necessary to abstract the daywork value from the daywork sheets, care being taken to compare like with like, i.e. materials not to be included in value, nett figures to be used if cost system is nett. This requires the daywork system to be constantly up to date, which is, in itself, a worthy bonus. This value is then compared in total with the allocated time spent on dayworks.

SUPERVISION

Supervision may be

1. direct site supervision—by, for example, gangers and foreman, or
2. over-all supervision—by, for example, agent, engineers, timekeepers.

It may be dealt with in two ways.

1. By allocating cost of supervision against individual operations. The standards or values must then include an element of supervision to balance this.
2. Supervision can be shown as a fixed labour overhead and a weekly standard shown in comparison. Staff can be included

in this item, although for confidential reasons staff salaries need not be included in the costing system, or can be included as an approximate lump sum so that individuals' salaries cannot be identified.

Direct site supervision lends itself better to allocation than does over-all supervision. However, the actual method used for dealing with supervision and the dividing line between direct and over-all supervision are not so important provided that consistency is maintained both within the cost statement and in feed-back of output data.

NATIONAL INCREASES

Where national increases are recoverable under the terms of the contract, they must be recorded separately for recovery purposes. The amount recorded is therefore both cost and value for national increases and can be shown as such under the heading 'Variable overheads', or can be listed with, for example, dayworks under a separate heading 'Recoverables'.

Where the contract is 'fixed price', an allowance will have been made at tender for possible increases. This allowance can be valued out each week as are other variable overheads, or can be phased, like redundancy payments, to occur only in the latter part of the contract.

Example

First 25% of contract: nil % on nett labour
Second 25% of contract: 4% on nett labour
Third 25% of contract: 8% on nett labour
Fourth 25% of contract: 12% on nett labour

That is, 6% average on total nett labour.

The cost of the national increases has to be recorded in the same way as are recoverable increases. *Figure 12.2* shows a typical summary sheet of recoverable items on a contract where selective employment tax has been approved by the Department of Employment and Productivity for a refund.

BONUS

In gross costing bonus is included in the average gross rate, but in nett costing bonus remains as a floating item that cannot be set against any known standard.

ANY FIRM & SON LTD.

RECOVERABLES WEEKLY SUMMARY

RECOVERABLES	DIRECT LABOUR	
FROM CLIENT	£ p	
BASIC RATE INCREASE		
EXTRAS OVER RATE		
TOOL MONEY		
LODGING ALLOWANCE		
HOLIDAY STAMPS		
STATE PENSION		
NATIONAL INSURANCE CONTRIBUTION		
NON-PRODUCTIVE OVERTIME		
MAINTENANCE TIME		
INCLEMENT WEATHER TIME		
GUARANTEED MAKE-UP		
TOTAL RECOVERABLE FROM CLIENT		

	DIRECT LABOUR	ALL STAFF
FROM GOVERNMENT D.E.P.	£ p	£ p
S.E.T. REFUNDS AND PREMIUMS		
REDUNDANCY REBATES		
TOTAL RECOVERABLE FROM D.E.P.		

CONTRACT _____   W/E _____

Figure 12.2

There are two types of bonus payment.

1. Attraction money: a policy, spot or standing bonus paid as an incentive to labour to work for a particular company or on a particular site. This kind of bonus is a variable overhead and may well have been allowed for in the calculation of overheads at tender. It should therefore be shown on the cost statement under variable overheads and compared with any value allowed.
2. A measured bonus system paid as an incentive to better production. This type of bonus attempts to improve outputs and reduce expenditure. The cost of the bonus scheme must therefore be offset by the savings on productive work. This can be done by two methods. (a) If the bonus is not of great significance, it can be shown as a cost with no value to all measured work, either by trades or in total, as illustrated in the standard cost example in Chapter 10 (*Figure 10.30*). (b) An enhanced average nett rate can be calculated each week to include nett wages plus bonus. Again, this can be done either by trades or in total.

CONCRETE MIXING

Normally the mixing, transporting and placing of concrete can be allocated to the appropriate measurable item of work, but on a large site with a central batching plant this may not be easy or management may wish to study the economics of mixing concrete as a separate operation. This can then be shown as a measurable item under the concrete trade, the standard for mix concrete having been abstracted from the mix, transport and place standards. Mixing mortar for brickwork, blockwork, etc., can be studied in a similar manner.

MAINTENANCE OR GREASING TIME

Maintenance or greasing time can either be dealt with as an overhead or charged against the appropriate measurable item. On a small contract with little plant a separate study of maintenance costs is not necessary. However, on a large contract with, perhaps, a fitting shop or maintenance depot such items as maintenance time are an important drain on resources and must be studied as a fixed labour overhead compared with any allowance made at tender for maintenance.

GENERAL ATTENDANCE

Theoretically all items of attendance actually out on site can be charged against some item of productive work—for example, picking up cut-offs from shuttering charged to the item of shuttering, cleaning spoil off road charged to the relevant excavation item, etc. Unfortunately this is not so easy in practice. Operations of tidying up site frequently involve a sudden purge or a gang of labourers doing nothing else but keeping a site clean and safe. It is then a difficult task to allocate times against different operations. The simplest solution is to make an allowance for 'General attendance' or 'Clean site' under the fixed overheads heading. General items of 'Transport on site', 'Unload materials', etc., can be similarly listed and thus prevent feed-back of output data of measurable items being contaminated by attendance items that may vary from site to site.

PIECEWORK

Productive work paid for on a piecework basis either to labour only sub-contractors or to direct employees has a known value or standard and a known actual cost. This work can be costed in the same manner as time-paid employees but by measurement of actual costs instead of labour allocation of costs, as illustrated in *Figure 12.3*.

DESCRIPTION	MEASURE m²	UNIT VALUE, £	TOTAL VALUE, £	UNIT COST, £	TOTAL COST, £	GAIN, £	LOSS, £
Formwork to bases	10	1.25	12.50	1.00	10.00	2.50	–
Formwork to columns	60	2.00	120.00	1.75	105.00	15.00	–
Formwork to beams	75	1.75	131.25	1.75	131.25	–	–
Formwork to ribs	4	2.50	10.00	3.00	12.00	–	2.00
Total carpenters			273.75		258.25	17.50	2.00

Figure 12.3

However, where a contract is almost wholly carried out on a piecework basis there seems little point in extending lists of known gains and losses each week simply to show a weekly summary. Office time can be better spent studying costs outside the piecework payments, e.g. dayworks, overheads, variations, etc. If a weekly summary is necessary, an average percentage gain or loss can be calculated for all piecework by trades or sections or in total as soon as rates are known and this percentage can be used to calculate values from the measured weekly payment sheets. For example, if average margin on carpenters for the whole contract is known to be 7% above cost and work carried out this week has cost £258·25, then value this week for carpenter is £258·25 + 7% = £276·33—a gain of £18·08, which is sufficiently accurate for general statistical use.

This method can also be used on incentive schemes based on work study, where payment is geared to output performance in such a way that payment for a standard hour of work is always the same no matter what rating men work at: for example, payment at 100 performance set at basic rate plus, say, 50%; therefore payment at 80 performance = (basic + 50%) × $\frac{80}{100}$. The cost per standard hour remains the same unless performances are outside the scale of payments: for example, payment at 60 performance would be below basic pay and this would have to be made up; the cost per hour would then rise. Similarly, if a ceiling is introduced, cost per standard hour above the ceiling would decrease.

COSTING BY STAGES

Certain types of repetitive work are suitable for grouping into stages of construction rather than units of measurement, e.g. housing or industrialised building. House building can be divided into stages of:

1. Excavate footings
2. Concrete footings
3. Brickwork up to D.P.C.
4. Hardcore fill
5. Concrete ground floor slab
6. First lift brickwork
7. Second lift brickwork
8. First floor joists
9. Third lift brickwork
10. Fourth lift brickwork
11. Top out brickwork
12. Roof joists
13. Floor boarding
14. Internal partitions
15. Joinery

In addition, there will be any other stages that the contractor intends to carry out himself rather than sub-contract. Drainage,

Costing Sundry Items

oversite excavation and external works are not easily grouped into such stages and may have to be measured as with a normal standard cost.

Allocation of labour and plant is simplified when made against one of these 15 stages. In practice some of these stages may be sub-let to labour only contractors and the payment per stage

HOUSE NO.	1	2	3	4	5	6	7	8	9	10	11	12
Excavate footings	√	√	√	√	√	√						
Concrete footings	√	√	√	√	√							
Brickwork to D.P.C.	√	√	√									
Hardcore fill	√											

Figure 12.4

previously agreed, which would simplify allocation even further. The standard or value for each stage can be calculated by previous measurement extended at normal unit standards. Weekly measurement can then be in units rather than square metres, etc., and can be recorded in a simple tick book, as illustrated in *Figure 12.4*.

For progress records the ticks can be recorded as a date or in a colour code. Quantities can be abstracted direct from such a tick book to a feed-back collection sheet, as shown in *Figure 12.5*.

CONTRACT _____

FEED-BACK RECORD SHEET

STANDARD VALUE	£25		£30		£40		£15		
ITEM	Excavate Footings		Concrete Footings		Brickwork to D.P.C.		Hardcore Fill		
WEEK NO.	No.	Hours	No.	Hours	No.	Hours	No.	Hours	
3	2	125							
4	3	184	3	220	1	93			
5	3	202	2	149	2	175	1	36	

Figure 12.5

STANDING TIME

Labour

Standing time for labour is best indicated as a cost to a particular section of measured work, e.g. trade or building with no value or standard set against it, thus causing the standing time to 'stick out like a sore thumb' in the loss column of the standard cost statement. It goes without saying that no attempt should be made to hide standing time under measurable work, as this defeats the object of the costing system.

Plant

A certain amount of plant standing is to be expected and can be allowed for, preferably as a plant overhead rather than as an inclusion in output standards. A budget of, say, 20% on all plant on site can then be calculated as the standard value for plant standing and all plant allocated as standing set against this. The difference between standard and actual is of great importance on, for instance, a motorway site. Plant standing time and plant maintenance time can be costed together on smaller sites where they are insignificant.

APPRENTICES, ETC.

Generally the youths or apprentices on site do not warrant special conditions regarding costs and can be treated as fully paid men for costing purposes. However, if this is found to be constantly disturbing the cost balance, such youths can be treated as being a proportion of a man related roughly to their wages scale.

ADJUSTMENTS

Although small mathematical errors or use of incorrect rates or measurement can be corrected on the next cost statement, it is preferable to show large adjustments as a separate statement; otherwise an error of, say, £100 one week, if corrected the following week, would cause an unrealistic increase or decrease of that £100 in the next cost statement. This may lead site management into research that is unnecessary or may induce a sense of improved profitability that is, in fact, only an adjustment. If feed-back is not being recorded and the cost is being used solely for site control, then adjustments are usually meaningless once the costs have been studied and appropriate action has been taken.

Costing Sundry Items

TO-DATE FIGURES

It is helpful to be able to watch the trend of some items, e.g. overheads, and for this reason the cost statement can be arranged to show the 'to-date' or 'last month' figures alongside this week's statement. Alternatively, key items can be plotted graphically either on a separate form or as an extension to the feed-back file, as illustrated in *Figure 12.6*.

GRAPHICAL CONTROLS

A wide variety of graphs, histograms and charts can be used to show pictorially the financial position of the various elements of a contract.

Figure 12.7 shows a histogram of values and costs for a contract set against the programmed value of work. This not only shows gains or losses but also provides an indication of general progress.

CONTRACT							
		FEED-BACK RECORD SHEET					
STANDARD VALUE	£0.50 = 1.25 Hours/m²				To Date	OUTPUT	WEEK
ITEM	Fixing soffit shutters including fixing props and span-forms				Standard 1.25 h/m² h/m² 3 2 1		
WEEK NO.		£0.40					
	m²	LAB. HOURS					
12	40	72					
	_40_	_72_					
13	123	150					
	_163_	_222_					
14	211	236					
	_374_	_458_					
15	194	228					
	_568_	_686_					
16	208	231					
	_776_	_917_					

Figure 12.6

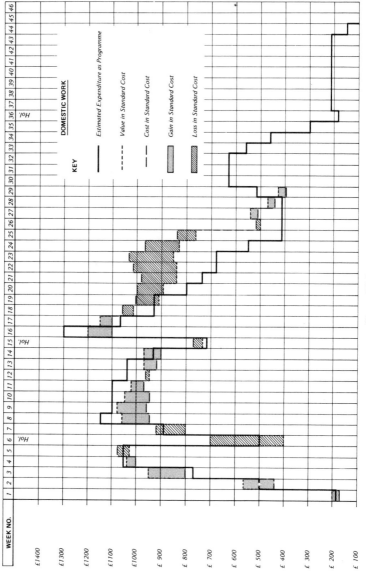

Figure 12.7

Costing Sundry Items

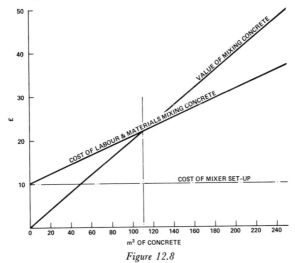

Figure 12.8

Such a histogram can be used to show all work or individual trade totals. Values and gains can be shown in one colour, costs and losses in another colour. Labour, plant and consumables can be shown together or as a separate chart. The decision on which items to show separately rests largely with the type of contract and likely trouble spots on the site. Another important factor being the reaction to quantitative detail by site management, where a standard cost statement may make little impression, a graphical presentation may have the desired effect.

Figure 12.8 shows a break-even chart used to calculate the average weekly concrete output required to break even on the cost of mixing concrete.

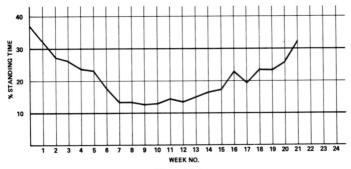

Figure 12.9

Figure 12.9 shows a maximum/minimum graph on which standing time for concrete transporters has been plotted. If the graph shows above the maximum line, management must check whether a transporter can be sent off site; likewise, below the minimum line management must check whether an additional transporter is needed.

Other suggestions for graphical records are:

1. Value or number of materials on site. An excess of materials on site leads to increased wastage, whereas too small a stockpile leads to hold-ups in programme.
2. Number of men on books each week compared with number programmed.
3. Percentage of absenteeism.
4. Ratio of staff salaries to measured work.
5. Percentage of pay offs each week.
6. Cubic metres of concrete placed each day or week; cubic metres of excavation or number of bricks, etc.
7. Wastage percentages of materials.
8. Comparison of trade gains or losses so that common changes in profitability can be related to weather, holidays, illness, changed conditions, etc.

EFFECT OF STANDARDS ON VARIABLE OVERHEADS

As the value of variable overheads varies directly with the value of measured work, it follows that an exceptional gain or an exceptional loss on measured work will affect the margin of gain or loss on the variable overheads, thus accentuating the results of the measured work on the overheads. *Figure 12.10* shows the final summary of measured work and fixed overheads, and of variable overheads where the total value of variable overheads adds up to 60% of the total value of measured work plus fixed overheads. A gain of £50 is shown on variable overheads with no losses. However, the reason

DESCRIPTION	VALUE, £	COST, £	GAIN, £	LOSS, £
Measured Work and Fixed Overheads	2000	1500	600	100
Variable Overheads	1200	1150	50	—

Figure 12.10

for this gain is not so much the economy on the variable overheads as the 25% gain on value of measured work and fixed overheads. If this 25% gain were to be removed and the value of measured work and fixed overheads taken at cost, i.e. £1 500, then the value of variable overheads would be much less, i.e. £1 500 × 60% = £900. Compared with this imaginary figure, the variable overheads would now show a loss of £1 150 less £900 = £250, which obviously requires further research and would have remained hidden if such an imaginary calculation had not been made. It is therefore advisable always to check how profitable on-costs would have been if the measured work were running at par.

13
MATERIALS

Although it is possible to allocate materials to individual operations and thus include a material element in any unit or standard costing system, the effort that has to be put into such an exercise does not make such a method practicable for regular costing and should therefore only be used when losses on materials cannot otherwise be identified.

There are two danger points that may cause a loss of money on materials, and a check on these two points is usually sufficient to control material losses.

BUYING MARGIN

Normally an estimator will have based his pricing of the material element in a bill rate on a quotation received from a local supplier of each type of material. However, owing to an unrecoverable increase in prices or to the lack of quotation at tender, the amount that the contractor has to pay for his material may eventually be more than he has allowed for. For example: at tender a contractor allows for using concrete aggregate from a local gravel quarry at a quotation of 75p/m^3; but when he commences construction, he finds that the aggregate is not up to specification and he has to purchase elsewhere at £1·00/m^3. He is therefore left with a 25p loss on every cubic metre of aggregate. If the estimated amount of aggregate in all concrete mixes for the contract is 10 000 m^3, it follows that a loss of £2 500 is inevitable on this material.

This may be offset by gains on other materials where, for instance, a post-tender quotation has shown a saving on the material element used by the estimators.

Such gains and losses on the buying margin of all materials can be tabulated as illustrated in *Figure 13.1* in order to calculate the over-all buying margin for a complete contract, the material element

ANY FIRM & CO. LTD.				CONTRACT		
MATERIAL BUYING MARGINS						

MATERIAL	VALUE ALLOWED PER UNIT £	ACTUAL COST PER UNIT £	GAIN PER UNIT £	TOTAL GAIN £	LOSS PER UNIT £	TOTAL LOSS £
5000 tonnes hardcore	0.60	0.45	0.15	750	—	—
600 tonnes cement	8.25	8.25	—	—	—	—
20 tonnes lime	10.10	9.90	0.20	4	—	—
100 tonnes building sand	0.65	0.70	—	—	0.05	5
1500 tonnes concreting sand	1.00	1.10	—	—	0.10	150
4000 tonnes aggregates	1.30	1.25	0.05	200	—	—
120 000 common bricks	10.00	9.90	0.10	12	—	—
30 000 engineering bricks	25.00	24.00	1.00	30	—	—
				996		155
				155		
		OVERALL BUYING MARGIN:		£841		

Figure 13.1

being the tender value for material excluding planned profit but including any margins allowed by the estimators for increased rates, etc.

Only if constant non-recoverable alterations to material prices continue throughout a contract is it necessary regularly to review this calculation. Reference to the buying margin must always be made before placing emergency orders with an alternative supplier owing to difficulty in obtaining materials, and the advantage of speeding up deliveries must be balanced with the increase in costs. Where the same material is being purchased from two or more suppliers, the cheaper supplier must be clearly marked so that daily or weekly requirements do not go to the dearer supplier simply because he arrives a little more quickly or provides a marginally better service.

WASTAGE

Inevitably a certain percentage of wastage must be incurred on any construction site owing to cut-offs in timber, snap headers and closers in brickwork, consolidation in hardcore, etc. Allowances will have been made at tender for such wastage but must be checked to ensure that normal limits are not being exceeded. The amount of each material purchased must be compared with the amount accounted for.

Figure 13.2 calculates the wastage for a sample four items of material based on the following information:

Deliveries to site

Bricks	53 000
100 mm dia. pipes	125 m
150 mm dia. pipes	420 m
Mesh reinforcement	210 m^2

Stock on site

Bricks	1 000
100 mm dia. pipes	30 m
150 mm dia. pipes	90 m
Mesh reinforcement	60 m^2

Amount in measured work

Brickwork 1B thick	358 m^2
100 mm dia. pipes	90 m
150 mm dia. pipes	300 m
Mesh reinforcement	140 m^2

CONTRACT												DATE up to 6/5/72	

| MATERIAL | UNIT | VALUE | | | ACTUAL | | | TO DATE | | | SINCE LAST CHECK | | |
		QUANTITY PLACED	WASTAGE ALLOWANCE	TOTAL ACCOUNTED FOR	BOUGHT	STOCK	USED	GAIN	LOSS	%	GAIN	LOSS	%
Bricks	No.	358 × 116 41 528	5% 2 075	43 603	53 000	1 000	52 000	—	8 400	20%			
100 mm dia. pipes	m	90	5% 5	95	125	30	95	—	—	—			
150 mm dia. pipes	m	300	15	315	420	90	330	—	15	5%			
Mesh reinforcement	m²	140	15% 21	161	210	60	150	11	—	8%			

Figure 13.2

Waste

Allow 5% planned waste on bricks and pipes
Allow 15% planned waste and laps on reinforcement

Care must be taken to allow for all purchases. The addition of delivery tickets is acceptable provided that strict control is observed over handing these tickets into the office or a goods received book is kept. Otherwise, invoices should be used for calculating deliveries, an allowance being made from delivery tickets for materials not yet invoiced. Where alternative suppliers are used in an emergency, care must be taken to include these suppliers even though their contribution may be only small.

Stock on site is sometimes difficult to calculate exactly but will, over a long period, prove insignificant. Stocks are, of course, more easily estimated when they are low: early Monday morning, after the site has worked for the weekend, may thus prove the ideal time to take stock of, for instance, concreting materials. Stacked bricks can be calculated at 500 per cubic metre downwards, depending on how loosely stacked they are, and cement silos can be measured more accurately when they are full. In order to reduce the effect of inaccurate estimates of stock, these checks are better calculated to date; weekly or monthly figures can then be deducted and shown relative to the 'to-date' results; this also helps to indicate trends of wastage during the course of the contract.

Records of temporary works on site should be included in order that all materials used can be allowed for: for example, concrete in base slab for site office, drain pipes to temporary toilets, etc.

It is often found that the method of calculating concrete mixes can be the cause of a gain or loss on concreting materials; if it is discovered that the rule of thumb method used at the estimating stage is inaccurate in practice, then the comparison for wastage purposes is best based on a realistic concrete mix. Any gains or losses caused by the estimator's method of assessing quantities of material in a concrete mix can be calculated as a variation of the buying margin. Otherwise, actual amounts used would be compared with unrealistic standards.

Exercise

Show how you would set up a system of control for wastage and buying margins of concreting materials. Assume that the estimate has been based on a rule of thumb method of calculating the mix, whereas on site concrete is weighbatched. Assume also that cheaper quotations have been obtained since tendering.

14
SUB-CONTRACTORS

LETTING MARGIN

As with materials, it is necessary to check buying margins or letting margins on domestic sub-contractors at the commencement of a contract. Nominated sub-contracts are selected by the client and cannot be varied by the contractor; thus such a comparison is unnecessary except as a service to the client in obtaining quotations from various nominated firms.

Figure 14.1 shows the comparison of domestic sub-contracts, which would not normally require to be revised during the course

ANY FIRM & SON LTD.   CONTRACT

SUB-CONTRACTORS LETTING MARGINS

TYPE OF WORK	VALUE OF SUB-CONTRACT IN TENDER £	ACTUAL SUB-CONTRACT £	GAIN £	LOSS £
Structural Steelwork	25 340	25 340	–	–
Plumber	10 961	10 400	561	–
Plasterer	4 210	4 530	–	320
Felt Roofer	2 200	1 950	250	–
Painter	1 083	1 083	–	–
Fencer	846	813	33	–

Overall Letting Margin

844
320
£524

Figure 14.1

of a contract unless a sub-contract was changed or the sub-contractor's quotation altered owing, for instance, to inability to achieve programme or alteration in conditions on site beyond the control of the sub-contractor.

ATTENDANCE

Few contracts are completed without the main contractor providing some service or attendance on the sub-contractor. This may be recoverable from the client either as a daywork or as a measurable item or may be recoverable from the sub-contractor (depending on the conditions of sub-contract) as a deduction from his account. Such attendance may, however, be the responsibility of the main contractor and as such must be included in the site costing system so that a regular comparison can be carried out with whatever allowances were made at tender. This can be done as a fixed labour overhead, the value being either a fixed sum per week or related to the amount of sub-contractor's work completed. As sub-contractors are usually valued only monthly, a conversion of a percentage allowance to a weekly allowance is necessary for a weekly costing system. Probably the simplest method of budgeting for attendance on sub-contractors is to allow a lump sum standard which is drawn off as the sub-contractor's work proceeds in a similar manner to that described in Chapter 7 in connection with temporary roads.

15
ACTION

Many facts shown up by the site costing system will, in themselves, suggest a possible remedy—loss on staff, over-cementing of concrete, high percentage of plant standing, etc. However, the majority of operations, overheads, etc., will still require further research. Whichever way a cost statement has been calculated, it is no more than a statement of *where* time and money have been expended and *where* losses have been made. Action must now be taken to find out *why* such losses have occurred and *what can be done* to put them right.

Reasons for losses can be summarised under the headings 'Site inefficiency', 'Interference by client' and 'Estimator's outputs'.

Site inefficiency. This may of course be due to incompetence, but is more likely to be the result of lack of communication, lack of planning, lack of materials, lack of suitable labour or lack of incentive. These possibilities must be studied and the root of the inefficiency discovered. Comparisons between gangs of men or similar sites may indicate a reason. Perhaps a bonus scheme is required; perhaps stores are not adequately controlled or lines of command not well defined. Pre-costing can help to set visible targets. Spot costs can make a more detailed study of a particular operation. Work study can isolate delays and interruptions and calculate method improvement.

It is helpful further to break down losses under this heading into, for example, the following categories:

1. Weather
2. Lack of continuity of work
3. Lack of materials
4. Excess labour on section
5. Making good
6. Quantities in excess of drawings

Interference by client. Many alterations, additions, deductions or changes may occur during the course of the contract, over which the contractor has no control. All such alterations must be charged to the client, as they were not envisaged in the estimate.

Losses must be checked to see whether they have occurred as a result of such an alteration and the machinery set in motion to recover excess costs either by: (*a*) daywork, (*b*) increased rate, (*c*) increased measure or (*d*) claim.

Estimator's outputs. Where the standard cost uses the estimate for calculation of standards, an error or misjudgement by the estimator may cause an incorrect standard, in which case a loss must be accepted. As feed-back of outputs to a standards library or to the estimator increases, so this kind of error should disappear.

ACTION SHEET

In order to obtain further information from gangers, foremen, etc., as to the reasons for losses, an action sheet is advisable similar to the example illustrated in *Figure 15.1*. The more important losses are thereby put to those closest to the operation and their comments either filled in by them or made verbally and entered in the 'Action taken' column by a timekeeper, cost clerk, etc. This then gives site management the lead necessary to categorise the losses and take measures to recover costs or prevent further losses of the kind.

SPOT COSTS

The detail shown in the standard costing system may not be sufficient to indicate the true reason for a loss on a particular operation or particular section of work. A special cost study or spot cost is therefore required of this item, splitting operations down into more detail: for example, tradesmen and labourers kept separate; temporary works such as barrow runs, templates, etc., kept separate; various elements of the operation kept separate—fetch materials, cut to size, position, fix, clear up. The spot cost needs to give full details of these cost elements on the spot cost statement and must not be summarised on to a separate form for presentation to management, as this would defeat its purpose. *Figure 15.2* illustrates a spot cost for an operation of concreting a plinth that is known to be additional to contract and for which no similar rate exists in the bill of quantities. The spot cost will assist the quantity surveyors to build up a rate for this item.

CONTRACT				DATE	
		ACTION SHEET			
DESCRIPTION OF WORK	VALUE £	COST £	LOSS £	ACTION TAKEN	
Concrete base to crane beam	9	30	21	This beam had to be concreted in phases to provide access for other contractors. Q.S. please note	
Shutter base to temperature measuring post	2	12	10	¼ mile from joiner's shop. Would be better recorded as daywork	
Cutting steel with burning gear	—	15	15	Making good. Domestic loss	
Excavate compressed air line trench	32	48	16	Should have been measured through existing road. Record note now prepared	
Backfill land drain with stone	70	105	35	Appears in order but method study being carried out to see if any improvement can be made	
Strip shutter from gate post base	4	12	8	Shutter props part buried by our own temporary access road. Domestic loss	

Figure 15.1

CONTRACT _____

DESCRIPTION _Construct Pump Plinth_   SHEET NO. _____

DATE	DESCRIPTION OF OPERATION	TYPE OF LABOUR	HOURS	RATE £	AMOUNT £	TYPE OF PLANT	HOURS	RATE £	AMOUNT £
6 July 70	Clean off area of plinth	lab.	2 x 4	0.3750	3.00	2T Comp.	4	1.25	5.00
	Bush hammer floor slab	"	2 x 4	0.3750	3.00	2T Comp.	4	1.25	5.00
	Make shutters	carp.	1 x 5	0.4396	2.20				
7 July 70	Fix shutters to plinth	"	2 x 2	0.4396	1.76	Mobile crane	2	0.85	1.70
	Prop, line and level shutters	"	2 x 3	0.4396	2.64				
	Make boxes for pockets	"	1 x 4	0.4396	1.76				
	Fix boxes to shutter sides	"	2 x 3	0.4396	2.64				
8 July 70	Hang boxes in plinth top	"	2 x 4	0.4396	3.52				
	Line and level boxes in plinth top	"	2 x 4	0.4396	3.52				
9 July 70	Place ready-mix concrete in plinth	lab.	4 x 3	0.3750	4.50	Poker vib.	3	0.25	0.75
	Joiner standby during concreting	carp.	1 x 2	0.4396	0.88	Mobile crane	3	0.85	2.55
	Trowelling	lab.	1 x 1	0.3750	0.37				
11 July 70	Strip shutters	carp.	2 x 1	0.4396	0.88	Mobile crane	1	0.85	0.85
	Strip boxes	"	2 x 3	0.4396	2.64				
	Rub up concrete	lab.	1 x 1	0.3750	0.37				
				Labour	33.68				
				Plant	15.85				15.85
				Total Cost	£49.53				

Figure 15.2

Action

Allocation of labour and plant would normally be more carefully controlled than under a standard/costing system because only a small section of work is being studied. Therefore personal attention can be given to the spot cost allocation.

Spot costs can also be used as a check on the standard costing system to prove the accuracy of the allocation system, measurements, bookkeeping, use of correct standards, etc.

Where no standard costing system is in use, a spot cost on selected items can be used to ascertain the profitability of key operations and thus provide some degree of control.

WORK STUDY

Work Measurement

Just as a spot cost can look at an operation in more detail than a standard cost, so work measurement can provide the ultimate in the study of an operation or of a certain item of plant being used. By use of work study techniques (B.S. 3138 : 1969, Section 31) a more consistent standard for an operation can be ascertained under known conditions and known methods, the degree of detail and synthesis depending directly on the resources put into obtaining these standards. A good example is the data produced by the Federation of Swedish Building Employers which lists each job of work on three schedules, blue sheet containing details of method, brown sheet containing time standards and yellow sheet giving a detailed job description. Preferred methods are marked accordingly. The Swedish system of payment by locally agreed piecework rates was rationalised in 1934 to form a national list and is now being superseded by rates based on this work study data. Problems of learning curves are still unsettled, requiring differing standards on, say, a site of 50 dwellings to a site of 5 000 dwellings, but the yardsticks provided are nevertheless adequately suitable for use in standard cost control.

Method Study

By use of the various systems of recording methods and devising new methods listed in B.S. 3138 : 1969, Part 2, improvement to methods or more economical balance of labour and machines may be obtained. The multiple activity chart (clause 21201) is probably the most useful aid on a construction site, there being many occasions when men and machines are working together and on examination can be rearranged to produce either more output or a reduced unit cost. Examples are: gangs excavating and laying drains, joiners fixing shutters by crane, number of lorries being loaded by machine.

PRE-COSTING

By setting standards against the contract's weekly programme and comparing the total of these standards with the probable wage bill for the same week it is possible to calculate the profitability of the following programme even if that programme is not a written one and exists only as the thoughts of the foreman or manager. Thus site management can visualise how essential it is to complete to programme. If the planned programme indicates a loss, it is not too late to study that programme together with those responsible for carrying out the work, so that foremen and gangers can appreciate how near the bone the programme is—how essential it is, for instance, that they achieve that extra pour of concrete, strip that shutter a day earlier or get that JCB off site by Wednesday at the latest. If the programmed loss is inevitable through reasons beyond the contractor's control, then revised rates can be agreed with the client while the revised work is being carried out so that the effect of alterations can be seen by both parties at first hand. This will reduce the amount of paperwork normally involved in recording such revisions.

Figure 15.3 illustrates the pre-cost for the site used in the standard cost example of Chapter 10; the labour standards are as illustrated in *Figures 10.20–10.27* with the following additions:

> 150 mm dia. S.G.W. pipes   £0·20/m
> 215 mm brickwork   £1·25/m²

The pre-cost would normally be calculated on the Friday prior to the week being studied; the number of pay-offs would therefore be known and any new starters would, no doubt, have already been told to report on the following Monday. Thus the last known wage bill can be revised to allow for the current week's pay-offs and new starters and for those during the week being studied. That is:

Week 7 wage sheet produced by Thursday of week 8.
Pre-cost for week 9 calculated on Friday of week 8.
Week 7 wage sheet adjusted for four new starters and two pay-offs for weeks 8 and 9 is approximately equal to wage sheet for week 9.

			PRE-COST				
VALUE						WEEK NO.	9

MEASURABLE WORK		@	LABOUR, £	@	PLANT, £	@	SUB-LET GROSS L & P, £
Exc. o/s.	20 m^3	0.40	8.00	0.50	10.00		
Exc. basement	800 m^3	0.08	64.00	0.10	80.00		
Exc. bases	50 m^3	0.50	25.00	0.20	10.00		
Exc. drains	20 m^3	0.50	10.00	0.50	10.00		
Blinding	7 m^3	2.00	14.00				
Concrete bases	15 m^3	1.50	22.50				
Concrete floor slab	10 m^3	2.25	22.50				
Concrete drains	3 m^3	1.80	5.40				
Concrete columns	4 m^3	2.60	10.40				
215mm bwk.	20 m^2	1.25	25.00				
150mm dia. pipes	25 m	0.20	5.00				
Cart away	890 m^3					0.15	133.50
			211.80		110.00		133.50
PRELIMINARIES							
LABOUR							
Chainman			10.00				
Unload matls.			5.00				
PLANT							
Conc. mixer					8.00		
Mortar mixer					4.00		
			226.80		122.00		133.50
+ ON-COSTS 53.37%			121.50				
GROSS LABOUR			348.30				
PLANT			122.00				
SUB-LET			133.50				
TOTAL GROSS VALUE			£603.80				

Figure 15.3

In the example the assumption has been made that the site will again work a six-day week except for the 22RB gang, who will work a seven-day week as before. Concrete and mortar mixers will be as week 7.

Like post-costing, pre-costing can be calculated nett or gross. However, as approximations have to be made in pre-costing and no allocation is available, it is generally sufficient to study only measurable work, the normal post-costing system being relied on to indicate any adverse trends in overheads. For illustration the following example has been calculated gross.

Programme for week 9

Excavate oversite	20 m^3
,, basement	800 m^3
,, bases	50 m^3
,, drains	20 m^3
Blind bases	4 m^3
,, drains	1 m^3
,, floor slab	2 m^3
Concrete bases	15 m^3
,, floor slab	10 m^3
,, columns	4 m^3
,, drain surround	3 m^3
215 mm brickwork to m/h	20 m^2
150 mm dia. S.G.W. pipes	25 m
Cart away (on lorry)	1 100 m^3

Cost

	£	
Total wages (week 7)	326·00	(for 11 men)
New starters 4 − 2 = 2		
$2 \times \dfrac{£326}{11}$	59·00	
Sub-let 1 100 @ 0·10	110·00	
Plant 10/7	8·00	
5/3½	3·00	
22 RB	90·00	
	£596·00	Gross cost

Action

Comparison

	£
Planned value	603·80
Estimated cost	596·00
Planned gain	£7·80 or 1·3%

On a site using considerable plant this comparison can be carried out comparing the various elements of work.

Labour (gross)

	£
Planned value	348·30
Estimated cost	385·00
Planned loss	£36·70

Plant

	£
Planned value	122·00
Estimated cost	101·00
Planned gain	£21·00

Sub-let

	£
Planned value	133·50
Estimated cost	110·00
Planned gain	£23·50

The example shows that although plant and labour only sub-contractors have been planned economically, domestic labour is planned at a loss.

ADAPTABILITY

A cumbersome, complicated, untidy system of costing can be most effective if action is taken on its results, yet the smoothest, most straightforward, neatest cost statement can prove useless if not heeded. This book provides a variety of approaches to costing systems on the basis that, like good clothes, a costing system must be tailor-made to fit a firm, and must be capable of variation to suit particular conditions. Just as a well-dressed man will wear something a little looser for gardening, so a good costing system will discard some of its finer points on a smaller contract, yet the underclothes, the base on which the system is built, must remain constant. Otherwise, interchange of staff, collation of data, comparison of problems become a maze of adjusted adjustments that lack the ring of confidence so essential to a good site cost control system.

Exercises

1. Daywork has been valued in the costing system of a complex town centre contract at 10% above cost but a thorough spot check for 1 week comparing allocated expenditure of labour, plant and materials for daywork operations with the amount of daywork submitted to the client gave the following results:

| Expenditure | £120 |
| Daywork submitted | £90 |

Therefore loss on daywork £30 or $33\frac{1}{3}\%$ of value. Cost statement was showing £120 as cost and £132 as value, i.e. a gain of £12. Apparently expenditure is being missed at some point in the daywork system. State what the danger points are and what action you would take to prevent this happening in future.

2. Calculate a nett pre-cost, i.e. excluding all variable overheads, for week 10 of the standard cost exercise contract. Assume labour and plant strength to be as week 9 and programme to be as follows:

Excavate basement	600 m³
,, bases	75 m³
,, drains	40 m³
Blind bases	6 m³
,, drains	2 m³
Concrete bases	20 m³
,, floor slab	5 m³
,, drain surround	6 m³
215 mm brickwork to m/h	30 m²
150 mm dia. S.G.W. pipes	50 m
Cart away	900 m³

INDEX

Accounts, 1, 5-10
  contract, see Contract accounts
  firm's, 1
Action on losses, 137-145
Action sheet, 138
Adaptability of a system, 144-145
Adjustments to costs, 124
Allocation sheets
  coded descriptions, 33
  completed by ganger, 30, 31, 67
  completed by other than ganger, 34
  lost time, 59
  overheads, 38
  overtime, 37
  plant, 35, 36
  standard descriptions, 31, 32
Allocation time, 34
Allowed time, 42, 50, 63
Apprentices, 124
Attendance
  on client, 52
  on sub-contractors, 136
  on tradesmen, 42, 121
Attraction money, 54

Balance, 26, 97, 98
Bank charges, 1
Barrow runs, 29
Basic times, 50
Bonus, costing, 118
Bonus increment, 50
Bonus payments, 86
Bonus schemes, see Incentives
Breakdown of estimator's rates, 45
British standard on time study techniques of work measurement, 42, 50, 141
Buying margin, 4, 130

Cash records, advantages and disadvantages of, 65
Chainboy, 52, 53, 89
Charts, 125
Clean office, 53
Clean public roads, 52

Clean site, 121
Client's effect, 44, 137
Coded descriptions, 33
Computer, 34
Concrete mixes, calculating, 134
Concrete mixing, 120, 127
Contentious items, 6
Contingency allowance, 50
Continuity, 137
Contract accounts, 1, 5-10
Cost(s) and costing
  adjustments, 124
  by stages, 122
  civil engineering, 26
  gross, 11
  gross cash, 11, 22
  in cash, 65
  in hours, 65
  labour, see Labour costs
  nett, 11
  of small items, 115
  plant, 3, see also Plant
  prime, see Prime costs
  site, 2
  spot, 4, 138-141
  standard, see Standard cost and costing
  statement, 23, 24, 94, 95
  summary, 25, 96
  sundry items, 115-129
  to date, 125
  tradesmen/labourers, 22
  unit, see Unit costing
Cost control, introduction to, 1-4
Cost reconciliation, 97
Credits, 116
Curing concrete, 29

Data bank, 42, 141
Daywork operations, 2, 117
Deviation from normal pricing, 4, 5
Dumpers; see Transport on site

Earnings, of contract, 2
  programmed, 7

Index

Earthmoving, bulk, 26, 39
Errors in tender, 44
  in workmanship, *see* Making good
Estimating, breakdown of rates, 45
  feel of the market, 43
  pricing, 42
  standards, 48
Excess overtime, *see* Non-productive overtime
Expenditure data, 27–39
Expenditure of contract, 3

Fares, 53
Fatigue allowance, 50
Federation of Swedish building employers, 141
Feedback records, 58–59, 86–93
Financial programme, 7
Fixed price contracts, 118
Fuel and oils, 3, 36, 38

Ganger's allocation sheets, 30
Graduated pension scheme, 53
Graphs, 125
Greasing time, *see* Maintenance time
Gross costing, 11, 37, 38
Gross labour rate, 22
Guaranteed make up (GMU), 57

Head Office overheads, 3, 4, 45
Histograms, 125
Historical outputs, 40, 58
Holiday stamps, 3, 53
House building, stages, 122

Importation of labour, 54
Incentives, 30, 34, 58, 61–64, 82, 99, 120, 122, *see also* Bonus
Inclement weather, 53, 137
Inefficiency, site, 137
Insurances
  general, 53
  national, 3, 53
Interference by client, 137
Interim payments (valuation), 1, 7
Internal company adjustments to tender, 43

Job card system, 30, 64

Labour, allocation of, 38, 123, 141, *see* Allocation
  small items of, 115
  standing time for, 124
Labour costs, 3, 27
Labour only sub-contractors, 65, 87
Labour rate, gross, 22
Learning curves, 141
Letting margin, 4, 135
Lodging allowance, 53
Losses, action on, 137–145
  on materials, 130
  reasons for, 137, 138

Maintenance time, 120, 124
Making good, 137
Market, feel of, 43
Materials, 130–134
  losses on, 130
  permanent, 3
  temporary, 3
Measured work, 2
Method study, 141
Minibus, 52, 53
Mixer set-up, 35, 38, 89, 90
Multiple activity chart, 141

National increases, 53, 118, 119
Nett costing, 11, 37–38
Nominated materials, 2
Nominated sub-contractors, *see* Sub-contractors
Non-productive overtime, 37, 54, 91

Offices, *see* Temporary buildings
On-costs, 54
Output, factors affecting, 40
Output data, feed-back of, 58–59
  historical, 40
Output standards, 58
Overheads, 21, 24, 26, 38
  Head Office, 3, 45
  nett costing, 26
  plant, 36, 38, 115
  site, 3, 37
    fixed, 52
    standards for, 52–57
    variable, 53, 128
  variable, 118
    effect of standards on, 128
  wage sheet, 37
Overtime allocation, 37

Index

Performance, 21, 59, 62, 99, 122
Personal needs allowance, 50
Personal observation, 34
Piecework, 3, 121–122, 141
Plant, allocation of, 123, 141
   charges, 3, 35
   hire, 35, 37
   maintenance, 52
   overheads, 36, 38, 115
   small items of, 115–116
   standing time for, 38, 124
Plus rates, 54
Policy allowance, 50
Policy bonus, 54
Pre-costing, 141–144
Preliminaries, 2, 43, 52
Premium time, *see* Non-productive overtime
Pricing, estimator's, 42
Prime costs, 1, 5–10
   breakdown, 8
   calculation, 9
   summary, 6
Production control, 59
Profit and loss account, 1
Profitability, 2
   calculation of, 141
Programmed earnings, 7

Rate refunds, 3
Recoverables, 53, 118, 119
Redundancy pay, 53
Relaxation allowances, 42, 50
Retention, 6

Scaffolding, 53
Scrap materials, 3
Selective employment tax, 3, 53
Sick pay, 53
Site costs, 2
Site inefficiency, 137
Site overheads, 37
   fixed, 52
   standards for, 52–57
   variable, 53
Small items, costs of, 115
   of plant, 52, 115–116
Spot costs, 4, 138–141
Staff requirements, 34, 52
Standard cost and costing, 18
   basic principles of, 65
   basis of, 18
   check on mathematical error, 97

Standard cost and costing—*continued*
   definition of, 18
   example, 65–98
   exercise, 99–114
   in civil engineering, 26
   performances, 62, 99
   typical statement, 22–24
Standard descriptions, 31
Standards, 40–51
   effect on variable overheads, 128
   for site overheads, 52–57
   incorrect, 138
   library of, 40, 58
   output, 58
   textbook, 48
   work study, 42, 50, 59, 62
Standing time, 37
   for labour, 124
   for plant, 124, 127
Statement, cost control, 23, 24, 94, 95
Stock on site, estimates of, 134
Stop-ends, 116
Sub-contractors
   attendance, 136
   domestic, 3, 135, 136
   labour only, 3, 65, 87
   nominated, 2, 135, 136
Subsistence (lodging allowance), 53
Summary, cost control, 25, 96
Sundry items, costing, 115–129
Supervision, 42, 117–118
Survey of site, 53

Task sheets, 27, 99
Taxation, 1, 3
Temporary buildings, 3, 52, 53, 89
Temporary roads, 3, 52, 53
Temporary services, 52
Tidy site, 121
Time study techniques of work measurement, British Standard on, 42
To-date figures, 125
Toilets, 52
Tower cranes, 35
Training levy, 53
Transport on site, 38, 121
Travelling expenses, 3
Travelling time
   on site, 42
   to site, 53
Trends, 125

Unit costing, 11–17
  as control tool, 15
  basis of, 11
  disadvantages, 13
  statement sample sheet, 13
  typical form showing cash results, 13
Unload materials, 89, 121
Unoccupied time, 42

Valuations, *see* Interim payments
Variation of price (V.O.P.), *see* Recoverables
Vibrators, 38

Wage sheet, 68
  overheads, 37
  reconciliation, 97
Wastage of materials, 132–134
Weighting of money in tender, 45

Wet time, *see* Inclement weather
Work content, 50
Work measurement, 141
  time study techniques of, British Standard on, 42
Work study
  British standard, 42, 50, 141
  contingency allowance, 50
  data bank, 42, 141
  fatigue allowance, 50
  method study, 141
  multiple activity chart, 141
  performance, 21, 59, 62, 99, 122
  personal needs allowance, 50
  policy allowance, 50
  relaxation allowances, 42, 50
  standards, *see* Standards
  travelling time, 42
  unoccupied time, 42
  work content, 50
  work measurement, 141